AF865631

Chloroplast Structure and Function

Edited by Muhammad Sarwar Khan

Published in London, United Kingdom

Chloroplast Structure and Function
http://dx.doi.org/10.5772/intechopen.107770
Edited by Muhammad Sarwar Khan

Contributors
Batta Kucheli, Blanca Flor Iglesias Figueroa, Brenda Julian Chávez, Carmen Daniela González Barriga, Chatti Khaled, Choulak Sarra, Edward Alexander Espinoza Sánchez, Efrida Basri, Faiz Ahmad Joyia, Ghulam Mustafa, Jinping Qin, Lijun Zhang, Luis Ignacio Siañez Estrada, Muhammad Sarwar Khan, Quintin Rascón Cruz, Rhouma-Chatti Soumaya, Saefudin, Sigifredo Arévalo Gallegos, Stephanie Solano Ornelas, Sugey Ramona Sinagawa Garcia, Tania Siqueiros Cendón, Xinyou Wang, Ying Liu, Zhengsheng Li

© The Editor(s) and the Author(s) 2024
The rights of the editor(s) and the author(s) have been asserted in accordance with the Copyright, Designs and Patents Act 1988. All rights to the book as a whole are reserved by INTECHOPEN LIMITED. The book as a whole (compilation) cannot be reproduced, distributed or used for commercial or non-commercial purposes without INTECHOPEN LIMITED's written permission. Enquiries concerning the use of the book should be directed to INTECHOPEN LIMITED rights and permissions department (permissions@intechopen.com).
Violations are liable to prosecution under the governing Copyright Law.

Individual chapters of this publication are distributed under the terms of the Creative Commons Attribution 3.0 Unported License which permits commercial use, distribution and reproduction of the individual chapters, provided the original author(s) and source publication are appropriately acknowledged. If so indicated, certain images may not be included under the Creative Commons license. In such cases users will need to obtain permission from the license holder to reproduce the material. More details and guidelines concerning content reuse and adaptation can be found at http://www.intechopen.com/copyright-policy.html.

Notice
Statements and opinions expressed in the chapters are these of the individual contributors and not necessarily those of the editors or publisher. No responsibility is accepted for the accuracy of information contained in the published chapters. The publisher assumes no responsibility for any damage or injury to persons or property arising out of the use of any materials, instructions, methods or ideas contained in the book.

First published in London, United Kingdom, 2024 by IntechOpen
IntechOpen is the global imprint of INTECHOPEN LIMITED, registered in England and Wales, registration number: 11086078, 167-169 Great Portland Street, London, W1W 5PF, United Kingdom

British Library Cataloguing-in-Publication Data
A catalogue record for this book is available from the British Library

Additional hard and PDF copies can be obtained from orders@intechopen.com

Chloroplast Structure and Function
Edited by Muhammad Sarwar Khan
p. cm.

This title is part of the Physiology Book Series, Volume 26
Topic: Plant Physiology
Series Editor: Tomasz Brzozowski
Topic Editor: Jen-Tsung Chen

Print ISBN 978-1-80356-384-8
Online ISBN 978-1-80356-385-5
eBook (PDF) ISBN 978-1-80356-620-7
ISSN 2631-8261

We are IntechOpen, the world's leading publisher of Open Access books

Built by scientists, for scientists

7,000+
Open access books available

187,000+
International authors and editors

205M+
Downloads

156
Countries delivered to

Our authors are among the
Top 1%
most cited scientists

12.2%
Contributors from top 500 universities

CLARIVATE ANALYTICS
BOOK CITATION INDEX
INDEXED

WEB OF SCIENCE™

Selection of our books indexed in the Book Citation Index
in Web of Science™ Core Collection (BKCI)

Interested in publishing with us?
Contact book.department@intechopen.com

Numbers displayed above are based on latest data collected.
For more information visit www.intechopen.com

IntechOpen Book Series
Physiology
Volume 26

Aims and Scope of the Series

Modern physiology requires a comprehensive understanding of the integration of tissues and organs throughout the mammalian body, including the cooperation between structure and function at the cellular and molecular levels governed by gene and protein expression. While a daunting task, learning is facilitated by identifying common and effective signaling pathways mediated by a variety of factors employed by nature to preserve and sustain homeostatic life. As a leading example, the cellular interaction between intracellular concentration of Ca+2 increases, and changes in plasma membrane potential is integral for coordinating blood flow, governing the exocytosis of neurotransmitters, and modulating gene expression and cell effector secretory functions. Furthermore, in this manner, understanding the systemic interaction between the cardiovascular and nervous systems has become more important than ever as human populations' life prolongation, aging and mechanisms of cellular oxidative signaling are utilised for sustaining life. Altogether, physiological research enables our identification of distinct and precise points of transition from health to the development of multimorbidity throughout the inevitable aging disorders (e.g., diabetes, hypertension, chronic kidney disease, heart failure, peptic ulcer, inflammatory bowel disease, age-related macular degeneration, cancer). With consideration of all organ systems (e.g., brain, heart, lung, gut, skeletal and smooth muscle, liver, pancreas, kidney, eye) and the interactions thereof, this Physiology Series will address the goals of resolving (1) Aging physiology and chronic disease progression (2) Examination of key cellular pathways as they relate to calcium, oxidative stress, and electrical signaling, and (3) How changes in plasma membrane produced by lipid peroxidation products can affect aging physiology, covering new research in the area of cell, human, plant and animal physiology.

Meet the Series Editor

Prof. Dr. Thomas Brzozowski works as a professor of Human Physiology and is currently a Chairman at the Department of Physiology and is V-Dean of the Medical Faculty at Jagiellonian University Medical College, Cracow, Poland. His primary area of interest is physiology and pathophysiology of the gastrointestinal (GI) tract, with a major focus on the mechanism of GI mucosal defense, protection, and ulcer healing. He was a postdoctoral NIH fellow at the University of California and the Gastroenterology VA Medical Center, Irvine, Long Beach, CA, USA, and at the Gastroenterology Clinics Erlangen-Nuremberg and Munster in Germany. He has published 290 original articles in some of the most prestigious scientific journals and seven book chapters on the pathophysiology of the GI tract, gastroprotection, ulcer healing, drug therapy of peptic ulcers, hormonal regulation of the gut, and inflammatory bowel disease.

Meet the Volume Editor

Muhammad Sarwar Khan is a renowned scientist with a Ph.D. from the University of Cambridge, UK. He has a strong background in agriculture, education, and biotechnology, having supervised more than 120 Ph.D. candidates and MPhil students. Dr. Khan has made significant contributions to agricultural biotechnology, including developing transgenic sugarcane, pioneering plastid transformation in rice and sugarcane, and knocking out genes from the chloroplast genome of higher plants. His current research interests include synthetic biology, whole-genome sequencing, pan-genomics, and developing cost-effective therapeutics and edible vaccines for animals. He has received numerous awards and honors and is currently Dean of the Faculty of Agriculture, University of Agriculture, Faisalabad, Pakistan. Dr. Khan has received prestigious national and international awards and honors and is a fellow of the Royal Society of Biology.

Contents

Preface

Chloroplasts, vital plant organelles, are being studied for potential applications in biotechnology, agriculture, biopharming, and renewable energy, leveraging advancements in genome sequencing technologies.

The book, *Chloroplast Structure and Function*, consists of seven chapters. In Chapter 1, "Introductory Chapter: Biology and Biotechnological Applications of Chloroplasts", Dr. Khan highlights the potential of green plant pigments in agriculture, biopharming, renewable energy, and beyond. Further, he demonstrates the essential role of the pigments in plant adaptation and response. In Chapter 2, Dr. Liu et al. explain the characteristics of the chloroplast genome of several plants on the Qinghai–Tibet Plateau. Further, they use the chloroplast genome in molecular evolution and phylogenetics to study plant species and their genetic information. In Chapter 3, Dr. Soumaya et al. confirm the domestication of date palms in the Arabian Peninsula and demonstrate an important gene flow with North African palm populations, considering chloroplast DNA as the best molecule for determining the evolutionary history of plant species. In Chapter 4, Dr. Espinoza Sánchez et al. explain the importance of plastome engineering, which offers potential for food security, drug production, and sustainable energy; however, challenges remain in expanding to other crops. In Chapter 5, Dr. Khan et al. explain chloroplast recycling and stress tolerance in plants. Further, they briefly describe how plastids regulate plant responses to stress by producing pigments, metabolites, and phytohormones, communicating with the nucleus, releasing nutrients, and degrading proteins, and improving stress tolerance and adaptation. In Chapter 6, Dr. Kucheli explains that guard cells, containing chloroplasts and stomata, regulate plant productivity by controlling stomatal pore, CO_2 uptake, and water loss. Further, overexpression of Rieske FeS protein improves quantum efficiency and growth rates. In Chapter 7, Drs. Saefudin and Basri highlight how chlorophyll and tannin production from forestry waste can boost selling-price capacity, reduce environmental impact, and provide economic opportunities for small businesses, enhancing fabric coloring, food safety, and health benefits.

Muhammad Sarwar Khan
Faculty of Agriculture,
University of Agriculture,
Faisalabad, Pakistan

Chapter 1

Introductory Chapter: Biology and Biotechnological Applications of Chloroplasts

Muhammad Sarwar Khan

1. Introduction

Chloroplasts are semi-autonomous, photosynthetic organelles that play an essential role in photosynthesis. Beyond their basic role in photosynthesis, these extraordinary structures are critical to a variety of important cellular processes, including the response to stress and the synthesis of various biomolecules including vitamins, lipids, amino acids, phytohormones, and products of secondary metabolism [1]. Chloroplast transformation presents numerous benefits, such as the precise integration of a transgene at a predetermined location within the plastome through homologous recombination. It avoids issues such as gene silencing and position effects, ensuring elevated levels of transgene expression due to the high ploidy plastome. Furthermore, maternal inheritance prevents transgenic leakage through pollen, making transplastomic GMOs environmentally friendly. Numerous valuable traits have been purposefully incorporated to enhance the agronomic performance of crop plants. The chloroplast transformation system has proven instrumental in the expression of industrial enzymes and therapeutic proteins, indicating it as a promising solution to address persistent issues surrounding food security, drug manufacturing, and sustainable energy generation [2, 3]. Chloroplasts have firmly established their capabilities by proficiently expressing transgenes, and packaging recombinant proteins, thus protecting them from cellular degradations. This makes it easier to produce highly useful proteins on a large scale. Various segments of the chloroplast genome have served as stable markers for identification and evolutionary research. The ability of chloroplasts to produce tannin and chlorophyll also provides a sustainable pathway for the synthesis of a variety of natural pigments [4]. This boosts not only the market value of these plants but also endows them with biologically beneficial and environment-friendly attributes. Additionally, chloroplasts interact with the nucleus and other organelles of the cell to organize the acquisition of essential molecules needed for their resistance to stressful environmental conditions. They act as sensitive environmental sentinels that effectively produce substances to mitigate stress and, particularly, induce the activation of genes that are encoded by the nucleus, to develop stress resistance (**Figure 1**).

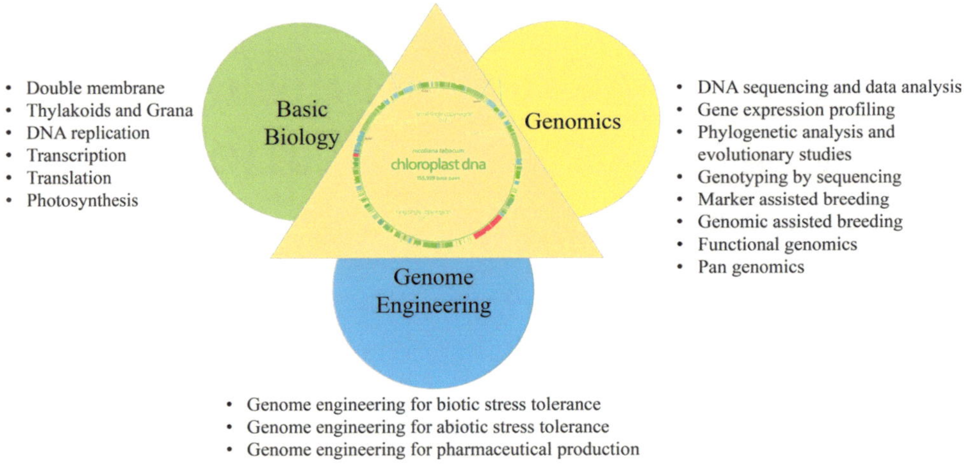

Figure 1.
Graphical representation of chloroplast biology and biotechnological interventions through plastome engineering.

2. Advances in chloroplast genomics

Plastid genomes, the 'plastomes' contain 120–130 genes and the majority of them are responsible for encoding critical parts of the organelle's gene-expression machinery and substances necessary for photosynthesis. These genes are neatly arranged within nucleoids. Tobacco was the first higher plant whose plastid genome was sequenced, opening the door for the sequencing and characterization of hundreds of new higher plants. Researchers now have access to an extensive amount of plastid genome data that will help them better understand how the genes in these plastids are functionally characterized. Many of the genes discovered in chloroplasts have undergone in-depth analysis to see how they affect the stability and metabolic functions of these organelles. A collection of genes called hypothetical chloroplast open reading frames (ycfs) is still unknown to have all of its specialized roles fully understood. Some of these ycfs have been classified as non-essential, while others have been classified as essential elements [5].

Recent advancements in sequencing technologies, such as next-generation sequencing (NGS), have made it possible to quickly and affordably conduct complete chloroplast genome sequencing. This is because chloroplast genomes contain an enormous amount of valuable information that is useful for species identification, phylogenetic inferences, and population genetic studies. Chloroplast genome analysis is more effective in understanding the complexities of progeny evolution and their evolutionary relationships. This decision is based on the haploid structure, maternal inheritance, and unusually well-preserved genes [6]. Furthermore, in recent times, chloroplast genomics has emerged as a highly promising tool extensively employed in phylogenetic investigations, owing to its exclusive maternal inheritance and absence of recombination events [7]. The study of chloroplast genetics has played a pivotal role in unraveling the dynamics of gene distribution, cytoplasmic diversity, and population divergence. Recently, chloroplasts have been associated with a number of vital functions within numerous plant processes, including plant defense mechanisms, the synthesis of defensive compounds, as well as the regulation of plant physiology, development, and the alternative splicing of transcripts [8, 9].

3. Plastids as a source of natural and engineered products

Innovative advances in the synthesis of chlorophyll and tannin have led to the development of their liquid, paste, and dry forms. This improvement in the production of natural pigments raises their market value while also promoting biologically favorable qualities that are in line with their sustainability and health advantages. Chlorophyll and tannin are abundantly available in wood waste, especially from forestry operations that use mangrove trees. These organic substances have the potential to replace synthetic pigments, in sustainable ways which are frequently connected to environmental issues. Utilizing wood waste in this way could promote and provide business opportunities in forest villages, mangrove industry hubs, and small- to medium-sized batik and weaving businesses. Batik artists who use coastal and inland motifs seem to prefer using natural chlorophyll/tannin hues. Therefore, research into different natural sources of tannin that are safe for food and drink as well as the human body is crucial for fabric and batik coloring.

Over the past few years, advances in chloroplast engineering have shown great promise in addressing critical global challenges such as ensuring enough food, producing medicine, and generating sustainable energy for our growing world population. Chloroplasts have proven to be effective at expressing transgenes and protecting the recombinant proteins from cellular processes, resulting in highly functional proteins. This characteristic has also been useful in the field of RNA interference technology. In addition to the practical advantages of chloroplast transformation, such as the elimination of positional effects, the capacity for polycistronic expression, and significant protein production, this method indicates a development in biosafety; however, even if its great biotechnological potential, crops that have efficiently transformed are still a proof of concept. Despite rigorous efforts, a few important crops have shown to be resistant to chloroplast transformation, which limits their ability to be grown more widely. This chapter covers the most recent developments in this field as well as the challenges that still need to be overcome before this method can be applied to more crops and become an important tool in the field of plant biotechnology.

4. Plastid engineering for crop improvement

The incorporation of specific genes into the chloroplast genome has proven effective in enhancing important agricultural qualities in crop plants. Due to its numerous benefits when compared to nuclear transformation, plastid modification has emerged as a powerful method for the genetic engineering of significant crops. Potato [10] and sugarcane [11] are the most recent examples. Many initiatives to increase food yields concentrated on improving plants' capacity for photosynthetic activity. Rubisco, the key enzyme at the center of photosynthesis, is made up of two parts: a large subunit encoded in chloroplasts and a tiny subunit encoded by the nucleus, which is later imported into the chloroplast. Similar to this, efforts have been made to change either the RuBisCO big subunit, the small subunit, or both parts [12] attempted to express complete RuBisCO protein in tobacco from *Synechococcus elongatus* and found CO_2 fixation rate and carboxylase activity of the RuBisCO to be increased. Another option for enhancing photosynthetic carbon fixation and agricultural yield is to increase the concentration of CO_2 in plastids. The introduction of the cyanobacterial bicarbonate transporter into the tobacco plastid genome failed to yield a significant enhancement in photosynthetic efficiency. However, when fructose-1, 6-sedoheptulose-1, and 7-bisphosphatase were expressed in both lettuce and tobacco chloroplasts, a notable boost in the productivity of these genetically modified plants became evident. The *chlB* gene from *Pinus thunbergii*, which is chloroplast-encoded, was also discovered to encourage tobacco root growth and early chlorophyll pigment development [13]. In order to increase biomass production, research is being done to convert C3 plants to C4 by modifying the RuBisCO large subunit and photo respirational pathway [14].

Since 1994, insect-resistant crops have flourished in fields, but Bt crops are raising concerns about the development of resistance. The development of insect-resistant transplastomic plants is a promising strategy, leveraging the benefits of transplastomic technology that have already been mentioned [10]. On the other hand, dsRNA expression, which targets an important insect gene in transplastomic plants, has been tested as a unique non-Bt-type insect resistance method. Within 5 days of feeding, RNA interference caused the target gene to be disrupted, killing all adult beetles and larvae [15]. When the agglutinin gene (pta) was expressed in leaf chloroplasts, aphids, lepidopteran insects, and bacterial and viral diseases were all successfully resisted [16]. The integration of the CeCPI gene (derived from sweet potatoes) and chitinase from *Paecilomyces javanicus* into tobacco plants conferred resistance not only to a wide range of pests and diseases but also rendered the plants resilient to salinity, osmotic stress, and oxidative challenges [17]. Proposing a potential defense against damage from excessive oxidation, increasing the activity of the mdar gene in tobacco plastids, and merging these chloroplasts with Petunia cells were suggested. This modification also seemed to make tobacco plants more resilient to oxidative stress. Additionally, when tobacco plants were genetically altered with the cyanobacterial flavodoxin (fld), they became better at handling oxidative stress. Furthermore, tobacco plants modified with the panD gene showed not only a 30–40% boost in their growth but also proved to be more resistant to high temperatures. Similarly, when arabitol dehydrogenase (ArDH) was introduced into tobacco chloroplasts, it allowed the plants to survive even in environments with extremely high salt concentrations, such as 400 mM NaCl [18]. This may open up novel opportunities for understanding how to cultivate resilient plants that can withstand stress and the value of transplastomic technologies.

5. Conclusion

In conclusion, chloroplasts are not only the most active cell organelle metabolically but also hold an immense potential for various biotechnological applications. Chloroplast being the main site of photosynthesis plays a pivotal role in a plant's responsiveness and mitigation to biotic as well as abiotic stresses. Studying its basic biological attributes helps to unfold new possibilities of innovations in agriculture, biopharming, renewable energy, and beyond.

Author details

Muhammad Sarwar Khan
Center of Agricultural Biochemistry and Biotechnology, Faculty of Agriculture, University of Agriculture, Faisalabad, Pakistan

*Address all correspondence to: sarwarkhan_40@hotmail.com

IntechOpen

© 2023 The Author(s). Licensee IntechOpen. This chapter is distributed under the terms of the Creative Commons Attribution License (http://creativecommons.org/licenses/by/3.0), which permits unrestricted use, distribution, and reproduction in any medium, provided the original work is properly cited.

References

[1] Song Y, Feng L, Alyafei MAM, Jaleel A, Ren M. Function of chloroplasts in plant stress responses. International Journal of Molecular Sciences. 2021;**22**(24):13464. DOI: 10.3390/ijms222413464

[2] De-la-Pena C, Leon P, Sharkey TD. Editorial: Chloroplast biotechnology for crop improvement. Frontiers in Plant Science. 2020;**13**:848034. DOI: 10.3389/fpls.2022.848034

[3] Yu Y, Yu PC, Chang WJ, Yu K, Lin CS. Plastid transformation: How does it work? Can it Be applied to crops? What can it offer? International Journal of Molecular Sciences. 2020;**21**(14):4854. DOI: 10.3390/ijms21144854

[4] Rascón-Cruz Q, González-Barriga CD, Iglesias-Figueroa BF, Trejo-Muñoz JS, Siqueiros-Cendón T, Sinagawa-García SR, et al. Plastid transformation: Advances and challenges for its implementation in agricultural crops. Electronic Journal of Biotechnology. 2021;**51**:95-109

[5] Khan MS, Riaz R, Majid M, Mehmood K, Mustafa G, Joyia FA. The tobacco chloroplast YCF4 gene is essential for transcriptional gene regulation and plants photoautotrophic growth. Frontiers in Plant Science. 2022;**13**:1014236. DOI: 10.3389/fpls.2022.1014236

[6] Khan AL, Asaf S, Lee IJ, Al-Harrasi A, Al-Rawahi A. First chloroplast genomics study of *Phoenix dactylifera* (var. Naghal and Khanezi): A comparative analysis. Plos One. 2018;**13**. DOI: 10.1371/journal.pone.0200104

[7] Liu H, He J, Ding C, Lyu R, Pei L, Cheng J, et al. Comparative analysis of complete chloroplast genomes of Anemoclema, Anemone, Pulsatilla, and Hepatica revealing structural variations among genera in tribe Anemoneae (Ranunculaceae). Frontiers in Plant Science. 2018;**9**:1097. DOI: 10.3389/fpls.2018.01097

[8] Singh NV, Patil PG, Sowjanya RP, Parashuram S, Natarajan P, Babu KD, et al. Chloroplast genome sequencing, comparative analysis, and discovery of unique cytoplasmic variants in pomegranate (*Punica granatum* L.). Frontiers in Genetics. 2021;**12**:704075. DOI: 10.3389/fgene.2021.704075

[9] Toufexi A, Duggan C, Pandey P, Savage Z, Segretin ME, Yuen LH, et al. Chloroplasts navigate towards the pathogen interface to counteract infection by the Irish potato famine pathogen. BioRxiv. 2019;**2019**:516443

[10] Hossain MJ, Bakhsh A, Joyia FA, Aksoy E, Gökçe NZÖ, Khan MS. Engineering of insecticidal hybrid gene into potato chloroplast genome exhibits promising control of Colorado potato beetle, *Leptinotarsa decemlineata* (Coleoptera: Chrysomelidae). Transgenic Research. 14 Sep 2023. DOI: 10.1007/s11248-023-00366-6

[11] Mustafa G, Khan MS. Transmission of engineered plastids in sugarcane, a C_4 monocotyledonous plant, reveals that sorting of preprogrammed progenitor cells produce heteroplasmy. Plants. 2021;**10**(1):26. DOI: 10.3390/plants10010026

[12] Lin MT, Occhialini A, Andralojc PJ, Parry MA, Hanson MR. A faster RuBisCO with potential to increase photosynthesis in crops. Nature. 2014;**513**:547-550. DOI: 10.1038/nature13776

[13] Nazir S, Khan MS. Chloroplast-encoded chlB gene from *Pinus thunbergii* promotes root and early chlorophyll pigment development in *Nicotiana tabaccum*. Molecular Biology Reports. 2012;**39**:10637-10646. DOI: 10.1007/s11033-012-1953-9

[14] Khan MS. Engineering photorespiration in chloroplasts: A novel strategy for increasing biomass production. Trends in Biotechnology. 2007;**25**:437-440

[15] Zhang J, Khan SA, Hasse C, Ruf S, Heckel DG, Bock R. Full crop protection from an insect pest by expression of long double-stranded RNAs in plastids. Science. 2015;**347**:991-994. DOI: 10.1126/science.1261680

[16] Jin S, Zhang X, Daniell H. Pinellia ternata agglutinin expression in chloroplasts confers broad spectrum resistance against aphid, whitefly, lepidopteran insects, bacterial and viral pathogens. Plant Biotechnology Journal. 2012;**10**:313-327. DOI: 10.1111/j.1467-7652.2011.00663.x

[17] Chen P-J, Senthilkumar R, Jane W-N, He Y, Tian Z, Yeh K-W. Transplastomic *Nicotiana benthamiana* plants expressing multiple defence genes encoding protease inhibitors and chitinase display broad-spectrum resistance against insects, pathogens and abiotic stresses. Plant Biotechnology Journal. 2014;**12**:503-515. DOI: 10.1111/pbi.12157

[18] Khan MS, Kanwal B, Nazir S. Metabolic engineering of the chloroplast genome reveals that the yeast *ArDH* gene confers enhanced tolerance to salinity and drought in plants. Frontiers in Plant Science. 2015;**6**:725. DOI: 10.3389/fpls.2015.00725

Chapter 2

Chloroplast Genome Characteristics of Plants on the Tibetan Plateau

Ying Liu, Jinping Qin, Zhengsheng Li, Lijun Zhang and Xinyou Wang

Abstract

Located in the interior of Asia, the Qinghai-Tibet Plateau is the largest plateau in China and the highest plateau in the world. It is also known as the "Roof of the world" and the "third pole". The various nature reserves on the Qinghai-Tibet Plateau are a treasure house of natural resources with the strangest ecological environment and the richest biological resources on the roof of the world. They are of high scientific value. This chapter will describe the chloroplast genome characteristics of several plants on the Qinghai-Tibet Plateau, such as *Aster, Asterothamnus centraliasiaticus, Aster altaicus, Corethrodendron multijugum*, *Clematis nannophylla*, and so on.

Keywords: chloroplast genome, *Aster*, *Asterothamnus centraliasiaticus*, *Aster altaicus*, *Corethrodendron multijugum*, *clematis nannophylla*

1. Introduction

Chloroplasts are important organelles in plants that help in photosynthesis [1]. Chloroplasts of higher plants are metabolic centers that sustain life on Earth through photosynthesis, as a semi-autonomous organelle with own DNA that encodes and has an independent genetic system [2]. For several reasons, cp genomes are more commonly used to study plant molecular evolution and phylogeny than mitochondrial genomes. The researchers note that differences in chloroplast genome sequences and their highly differentiated regions between plant species and individual plants not only enable a more comprehensive study of the taxonomy and phylogeny of these species, but also facilitate the identification, breeding, and conservation of valuable biological genetic information among related species [3, 4]. Here, we collected representative plants of alpine meadow, alpine steppe, desert steppe grassland types on the Qinghai-Tibet Plateau and determined their chloroplast genome sequences to understand their genome characteristics and phylogenetic evolution.

2. Species morphological characteristics and distribution information of five *Aster* species

Aster is a perennial herb of the Compositae family. Most *Aster* plants have well-developed roots, rhizoid type, and rapid reproduction [5]. With large root surface area, a large number of nutrients in soil can be absorbed and utilized for ecological restoration, giving full play to their ornamental and ecological value. It is vigorously used in greening and beautifying the environment to give full play to its ecological functions [6].

Aster has a flower head, with ligulate of purple or bluish-purple flowers and tubular of yellow flowers, which is of high ornamental value. *Aster* plants is one of the traditional Chinese herbs, and most *Aster* species contain a large number of important medicinal components such as flavonoids and quercetin [7–9], and the root and stem of *Aster* plants are commonly used as medicines. *Aster* species has a long history of medicinal use, there are many varieties of *Aster* distributed on the Qinghai-Tibet Plateau, and it is used in traditional Tibetan medicine as a medolome to treat seasonal epidemic diseases and clear heat and detoxification [10–12].

China is the main origin and distribution center of Eurasian *Aster* [12, 13], with about 123 species, including 82 endemic species. *Aster* has many species and abundant resources in China, widely distributed in northeast, northwest, southwest, and south China [5]. The Qinghai-Tibet Plateau has a unique geographical location, rich grassland resources, and high species diversity; according to the Flora of China, the Qinghai-Tibet Plateau is also one of the distribution centers of *Aster*. Here, we selected Guoluo Prefecture, the source of the Yellow River in the hinterland of the Qinghai-Tibet Plateau, to conduct a comprehensive survey of *Aster* germplasm resources and investigate the morphological characteristics of *Aster* species and determine the chloroplast genome of *Aster* species. Combined molecular tools with classical morphological information could prove a valuable information for a more accurate classification of Aster species.

2.1 Morphological characteristics of five species of *Aster*

2.1.1 Aster yunnanensis *var.* labrangensis *(Hand. -Mazz.) ling*

A. yunnanensis var. *labrangensis* is usually perennial herbs, and its height is about 20-70 cm (**Figure 1A**); Rhizome; stems erect and upper branched, stems 2–5 branched; both surfaces of leaves hairy and glandular, half clasped; Involucres hemispheric, 4-7 cm in diameter. The Ray florets are bluish-purple, the disk florets are yellow, the sparsely to moderately villous are 2 layers, short outside and long inside. Flowering period is July-August, and fruit period September-October. Generally found in hillside meadows, alpine meadow or thicket, elevation 3300–4300 m.

2.1.2 Aster farreri *W. W. Sm. et J. F. Jeffr*

A. farreri is usually perennial herbs, its height is about 30–60 cm (**Figure 1B**); rhizome long; stems erect, simple, sparsely to moderately villous. The leaves are basal and cauline, and sparsely villous, eglandular, margin entire or sparsely serrulate, villous-ciliate;blade oblanceolate to narrowly oblanceolate, 1.5–22 × 0.7–2.3 cm, leaf base gradually narrow, apex acute; middle cauline leaves linear-lanceolate, 7–13 x 0.7–2 cm, base rounded, subvalvate to valvate, apex acuminate; upper blade linear,

Figure 1.
Morphology and habitat map of five species of Aster. *(A)* Aster yunnanensis *var.* labrangensis*; (B)* Aster farreri*; (C)* Aster souliei*; (D)* Aster asteroids*; (E)* Aster poliothamnus.

about 2 x 0.1 cm, apex sharp. The capitulum is terminal, solitary, 5–8 cm in diameter, and the involucre is hemispherical, 2–2.4 cm in diameter; ray florets are purplish blue or lilac, disk florets are yellow; Pappus 2 layers, white or smudge white, outer layer very short. Flowering period is in July-August, and fruit period is in August-September. Generally distributed in Alpine, subalpine slope grassland, sand and gravel, elevation 2600–4200 m.

2.1.3 Aster souliei *Franch*

A. Souliei is usually perennial herbs, its height is about 2–45 cm (**Figure 1C**), sometimes caespitose; rhizomes robust, woody. Stems solitary, erect. The leaves are basal and cauline, the leaf surface is smooth to sparsely hairy, the margin is whole, and the basal leaves present at flowering stage are spatulate or obovate to oblanceolate; inflorescence terminal and solitary, 3–6 cm in diameter, involucral hemispherical, 6–8 mm; floret 25–55 radiate, bluish-purple to purple, tube glabrous, lamina 12–25 × 2–3 mm, glabrous, glandular; disk florets yellow,3–5 mm, pappus 1 layer, purplish brown. The flowering period is from July to August, and the fruit period is from August to September. It is generally distributed in alpine thickets, meadow, and hillside meadows, with an altitude of 2700–4500 m.

2.1.4 Aster asteroids *(DC.) O. Kuntze*

A. asteroids is usually perennial herbs, its height is about 2–25 cm (**Figure 1D**); rhizomes short; roots tuberoid, near ground surface, with 2–6 tuberous roots, radish shape, stems erect, solitary, 2–15 cm tall to 30 cm. Basal leaves are dense and live at flowering stage, obovate, or oblong, 1–4 cm long, 0.4–0.8 thin, 1.7 cm wide, leaf base tapered into a short stalk, subentire, with few fine teeth; middle leaf oblong or oblong spatulate, apex obtuse or acuminate, sessile, upper leaf linear. All leaves are with long hairs that are sparse or dense above, glabrous below or hairy only along veins, with long marginal hairs, with three veins off base. The capitula terminal are solitary at the end of the stem and are 2–3.5 cm in diameter. Involucral hemispherical, 0.7–1.5 cm in diameter; involucral bracts 2–3 layers, subequal, linear-lanceolate, 5.5–7 mm long, 1–1.5 mm wide, apical acuminate, abaxially and margin with purplish-brown dense hairs. Ray florets1 layer, about 30–60, tonguc bluish-purple, 10–20 mm long, 1–2 mm wide, apical point; disk florets orange-yellow, tube 1 mm long, split 1–2 mm long, with black or colorless glandular hairs; Achenes oblong, up to 3 mm long, white sparsely hairy or silky. The flowering period is from June to July, and the fruit period is from July to August. It is generally distributed in Alpine meadow, thickets, and hillside grassland, with an altitude of 2750–4800 m.

2.1.5 Aster poliothamnus *diels*

A. poliothamnus is usually perennial subshrubs, and its height is about 15–100 cm (**Figure 1E**), sometimes shrublike, caespitose, caudex woody. Stems branched, branches erect, there are dense leaves. Lower leaf withered; middle leaf oblong or linear oblong, about 1–2 cm long and 0.2–0.5 cm wide, entire margin, blade base slightly narrow or sharply narrow, apex blunt or pointed, leaf margin flat or slightly reversed; upper leaf small, elliptic. All leaves are short strigose above, pilose below, glandular on both sides, midvein raised below, lateral vein not visible. Inflorescence densely corymbose or solitary at branch ends; inflorescence peels are thin, 1–2.5 cm long, with sparse bracteoles. The involucre is broad campanulate, 5–7 mm long, 5–7 mm in diameter; involucral bracts 4–5 layers, imbricate arranged, outer oval or oblong-lanceolate, 2–3 mm long, all or upper grassy, apically apical, outer or only along midvein densely puberulent and glandular; inner layer up to 7 mm long, 0.7 mm wide, subleathery, upper grassy and reddish purple, marginal hairs. 10–20 Ray florets, lilac, oblong, 7–10 mm long, 1.2–2 mm wide. The disk florets are yellow, 5–6 mm long and 1.6–2 mm long. The crest is dirty white and about 5 mm long. Achenes oblong, 2–2.5 mm long, often ribbed on one side, covered with white silky hair. The flowering period is from July to August, and the fruit period is from August to September. It is generally distributed in arid slopes, alpine timberline, glades, and so on, with an altitude of 1800–3800 m.

2.2 Chloroplast genome characteristics of *Aster* species

We sequenced the chloroplast genome of five *Aster* leaves collected from Maqin County, Guoluo Prefecture, Qinghai Province, China (36°010 N, 103°450 E, altitude 3970 m). The chloroplast genome was sequenced. And the assembled chloroplast genome and its detailed annotation were submitted to GenBank (*Aster yunnanensis* var. *labrangensis*: OQ569735.1; *Aster diplostephioides*: OQ603807.1; *Aster farreri*: OQ603808.1; *Aster souliei*: NC_073537.1; *Aster asteroides*: NC_073536.1; *Aster poliothamnus*: OQ658689.1).

The complete chloroplast genome sequence lengths varied from 152,549 to 153,087 bp of five species of Aster (**Figure 2**), which has the quadripartite structure typical for most higher plants and highly conserved, respectively, which was divided into a LSC (84218-84,742 bp) and an SSC (18165–18,307 bp) regions separated by a pair of inverted repeats (IR, 24960–25,031 bp). In addition, the overall GC content of the chloroplast genome sequence was 37.3%. Since all the rRNAs were located in the IR

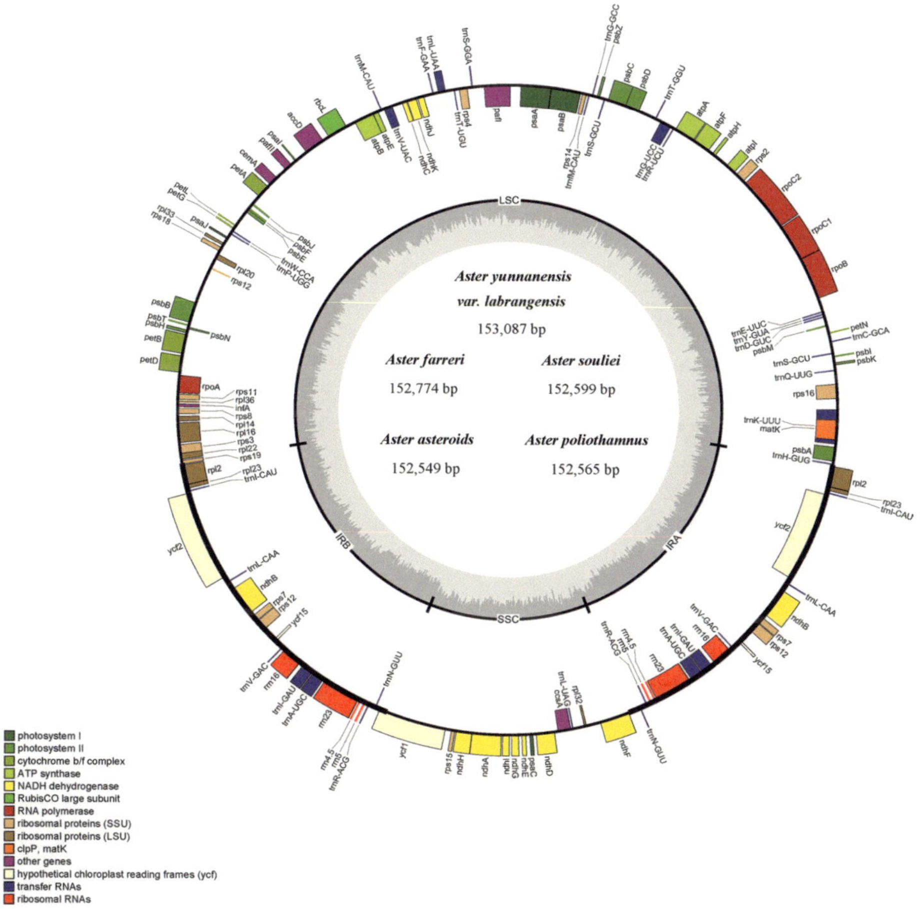

Figure 2.
Chloroplast genome map of five Aster *species. Genes inside the circle are transcribed clockwise, and those outsides are transcribed counter-clockwise. Genes of different functions are color-coded. The darker gray in the inner circle shows the GC content, while the lighter gray shows the AT content. The red and blue lines indicate GC skew, the red means GC skew greater than zero, while the blue means GC skew smaller than zero. Same as the chloroplast genome characterization below.*

regions, the GC content of the IR regions (42.9%) was higher than that of the LSC (35.2%) and SSC (31.2%) regions, respectively. A total of 130 genes were annotated successfully, including 85 protein-coding genes, 37 tRNAs, 8 rRNAs, respectively.

The CDS and tRNA of the five *Aster* species were mainly distributed in the LSC region, rRNA was only distributed in the IR region, and only one tRNA was distributed in the SSC region. Nineteen genes had two copies, which comprised of seven CDS coding genes, seven tRNA genes, and all four rRNA species (rrn16, rrn23, rrn4.5, and rrn5). In the genome, eighteen genes contain introns, among them 17 contains 1 intron and pafI contains 2 introns. The start codon of all CDS is ATG except rps19, where the start codon of rps19 is GTG, respectively.

Similarity in codon usage and amino acid frequencies were observed in five *Aster* species. And the results showed that about twenty-five thousand amino acids were detected in the chloroplast genome of five *Aster* species, in which Leucine was the most abundant, with two thousand and seven hundred codons (10.8%), followed by Isoleucine with two thousand and one hundred codons (8.4%), Serine and Glycine, with about 1960 and 1750 codons (7.7% and 6.9%), and Cysteine was the least abundant. There are 29 codons with RSCU values greater than 1 in five Aster. Methionine and Tryptophan had RSCU values equal to 1, but the most preferred codon was TTA, encoding Leucine (Leu), with an RSCU value of about 1.85.

SSRs are highly polymorphic molecular genetic markers, widely used, especially for population, evolutionary, and conservation genetics studies and forensics [14]. SSRs are composed of one or a few consecutive repeated nucleotides. By analysis of SSR dynamics in chloroplast genomes of five *Aster* species show that the pentanucleotide SSRs was only found in *A. asteroids* and *A. poliothamnus*, while the types and numbers of SSRs varied across species of *Aster*, indicating genetic diversity among species. A6 contains less mononucleotides and trinucleotides than the other four species of *Aster*. In five *Aster* species chloroplast genome, mononucleotide repeats are the most abundant repeats, followed by tetranucleotides, dinucleotides, and trinucleotides repeat, while pentanucleotides and hexanucleotides repeats rarely occur, and all of the mononucleotide repeats consisted of either A or T bases. SSR repeats in LSC region were much higher than SSC region and IR region in five *Aster* species.

The chloroplast genomes are of great significance in the reconstruction of plants phylogenetic relationships and evolutionary history [15]. In our study, we constructed a phylogenetic tree using the sequences of the whole chloroplast genomes of 41 species in the family *Aster*, including 24 (including *Heteropappus* and *Symphyotrichum*) species and using 15 species in *Senecio, Diplostephium*, *Asterothamnus, Artemisia*, and *Erigeron* as outgroups. The alignment of plastomes was generated by MAFFT. The maximum-likelihood (ML) analysis was performed using MEGA11, of which the bootstrap values were calculated using 1000 replicates with the best selected GTR + G model [14]. The result showed that all clades were strongly supported, *A. ageratoides* var. *scaberulus* is sister to a clade formed by *A. yunnanensis* var. *labrangensis*, *A. farreri*, *A. asteroids*, *A. souliei*, and *A. batangensis* according to the current sampling extent, and *A. poliothamnus* was sister to *A. sampsonii*.

2.3 Discussion and conclusion

In this chapter, the architecture of the basic characteristics, codon usage bias, SSRs and phylogenetic relationships of chloroplast genomes of five *Aster* species are studied. Five species of *Aster* have a typical quadripartite structure and 130 functional genes have been annotated. Moreover, in the coding strand, the bias of T over A existed in all five Aster species, while the bias of G over C differed. These findings, in combination with identified introns, codon usage bias, and SSRs, enrich our knowledge on chloroplast biology and genetic diversity of five *Aster* species and lay a strong foundation for further studies on molecular marker development, phylogenetic analysis, population studies, and chloroplast genome engineering.

3. Morphological characteristics and chloroplast genome analysis of *Aster altaicus* Willd

3.1 Species morphological characteristics and distribution information

Aster altaicus Willd., depicted in **Figure 3** [16], is a perennial herbaceous plant. It was previously classified as *Heteropappus altaicus* when it was part of the *Heteropappus* genus. However, due to its unique feature of ray floret having very short crown hairs, it was often associated with *Aster*, although some other *Aster* species also exhibit this trait. Eventually, it was reclassified solely under the *Aster* category [17]. Typically, these plants grow to be 15–40 cm tall and possess transversely or vertically woody roots. The stems originate at the base and give rise to multiple branches, which can be either erect or obliquely branched. These branches are covered with upwardly curved or spreading hairs and contain glands situated distally. The leaf blade is linear, oblong, or oblanceolate, measuring 3–35 mm in length and 1–7 mm in width. The anterior ends of the leaves can be blunt or acute and are usually complete, featuring short coarse or fine hairs on both upper and lower surfaces, accompanied by glandular dots. At the uppermost parts of the branches, multiple heads grow either individually or in an umbellate fashion. The involucre, approximately 1–1.5 cm in diameter, takes the shape of a hemispherical

Figure 3.
Morphology and habitat map of Aster altaicus. *From Yueliang Bay park in Guide County, Hainan Tibetan autonomous prefecture, Qinghai Province, China (101°43′95" E, 36°04′61" N), photograph by Ying Liu.*

structure and consists of 2–3 layers of nearly equal-length involucral bracts. These bracts are oblong-lanceolate or linear, approximately 5 mm long and 1.5 mm wide, with tapered pointed tips. They exhibit narrow membranous margins and glandular hairs on the abaxial side. The flower head contains 15–20 ray florets with tubes measuring around 2.5 mm. The ligules of these florets are blue, linear-oblong, reaching a length of 15 mm and a width of approximately 2 mm. The ray florets themselves are yellow, about 5 mm long, featuring five lobes of varying lengths and small external hairs. The achenes are oblong-obovate and hairy, with reddish-brown crown hairs, approximately 4 mm long and thicker. Flowering and fruiting occur from July to October [18].

A. altaicus Willd. thrives in various environments such as grasslands, deserts, sands, and arid mountainous areas at altitudes ranging from 0 to 4000 meters. This plant is widely spread in northern China and is mainly produced in Asia [17]. It is an important medicinal plant, widely used in traditional Chinese medicine and Mongolian medicine [19], and holds some forage value [20].

3.2 Characterization and phylogenetic appreciation of the chloroplast genome of *A. altaicus*

The chloroplast genome of *A. altaicus* leaves, collected from Yueliang Bay Park in Guide County, Hainan Tibetan Autonomous Prefecture, Qinghai Province, China (101°43′95“ E, 36°04′61” N), was sequenced. Subsequently, the assembled chloroplast genome, along with detailed annotations, was submitted to GenBank under accession number NC072176. The analysis revealed that the *A. altaicus* chloroplast genome is 152,473 bp in size and possesses an average GC content of 37.3%. It exhibits a typical four-part structure, including a large single-copy region (LSC) spanning 84,235 bp, a small single-copy region (SSC) spanning 18,218 bp, and two inverted repeat regions (IRs) spanning 25,013 bp (**Figure 4**). A total of 129 genes were successfully annotated, encompassing 85 protein-coding genes, 8 ribosomal RNA genes, and 36 transfer RNA genes. Among these genes, 17 contain one intron, while 3 contain two introns (**Figures 3** and **4**).

In this study, a phylogenetic analysis was conducted using the complete chloroplast genome of *A. altaicus* and 25 other *Aster* species, and two *Medicago* species (*Medicago monspeliaca* and *Medicago monspeliaca*) of Fabaceae (Two *Medicago* species were used as outgroups). The results showed that *A. altaicus* and *Aster altaicus* var. *uchiyamae* have a strong sister relationship.

3.3 Discussion and conclusion

A. altaicus Willd. belongs to the *Aster* genus of the *Asteraceae* family. Previous studies have explored the complete chloroplast genomes of various *Aster* species [21–25]. In this study, we present the first successfully assembled and annotated complete chloroplast genome of *A. altaicus* Willd., collected from the Tibetan Plateau. With a length of 152,473 bp, its sequence exhibits a typical tetramerization structure observed in other *Aster* species. Additionally, for systematic evolutionary tree analysis, we incorporated chloroplast whole genome data from 25 other *Aster* species. The results indicate a strong relationship between *A. altaicus* Willd. and *A. altaicus* from Korea, as well as *A. altaicus var. uchiyamae* and the samples used in this study. This research contributes to valuable genetic resources for future investigations on *A. altaicus* Willd. and proves to be essential in studying the phylogenetic relationships of *A. altaicus* Willd. It is crucial to continue exploring the genetic

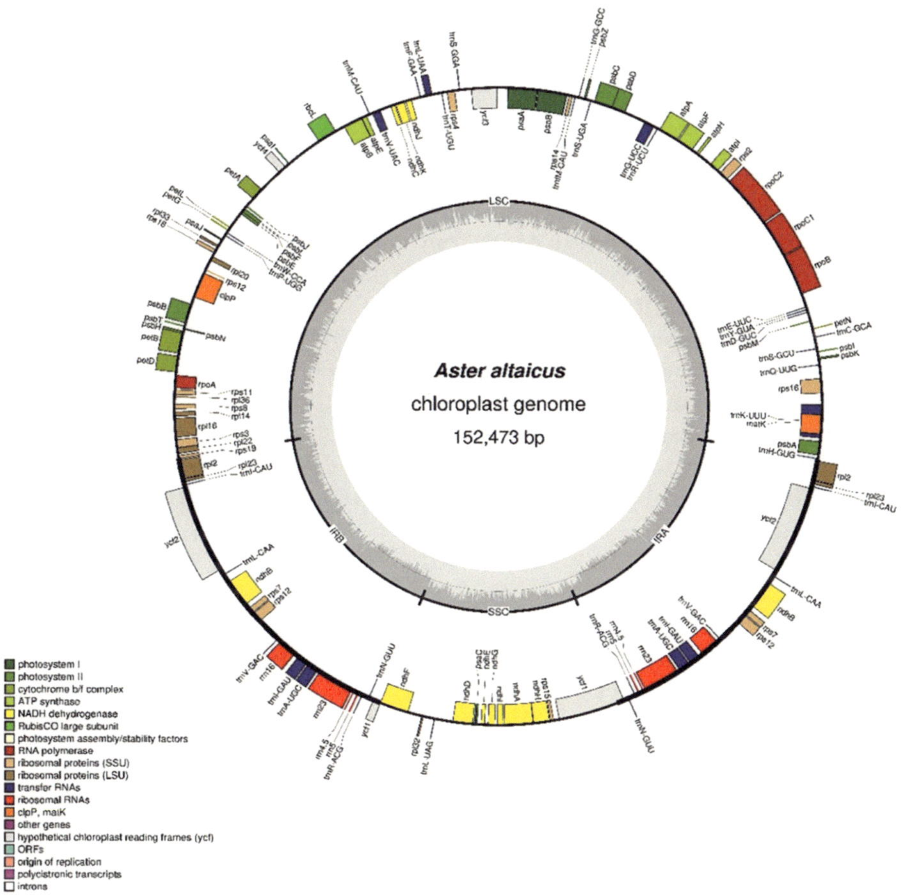

Figure 4.
Chloroplast genome map of Aster altaicus.

information of *Asteraceae* and advancing classification research within the *Asteraceae* family, including the *Aster* genus.

4. Morphological characteristics and chloroplast genome analysis of *Asterothamnus centraliasiaticus*

4.1 Species morphological characteristics and distribution information

Asterothamnus centraliasiaticus Novopokr. [26, 27] (**Figure 1**), also known as Aster centraliasiaticus and Aster alyssoides, is a perennial deciduous half-shrub of the family Asteraceae [28], native to arid and semi-arid areas of the Qinghai-Tibetan Plateau and northwestern Mongolia in China. It is a perennial deciduous half-shrub of the family Asteraceae [28], native to the arid and semi-arid regions of the Qinghai-Tibetan Plateau and northwestern Mongolia in China, and naturally distributed in the areas of 1300–3900 meters above sea level. *A. centraliasiaticus* is an ecologically important species in its range, especially for its role in soil stabilization and as a food resource for livestock and wildlife. In addition, *A. centraliasiaticus* has brightly colored flowers and can also be used as an ornamental plant. *A. centraliasiaticus* is a much-branched half-shrub, with plants up to 120 cm tall. Its stems are cespitose, branched below, with inflorescence branches above. *A. centraliasiaticus* is a not very tall half-shrub with many branches below, and a root neck buried in the soil, from which branches emanate. Its axial roots are deeply embedded in the soil up to about 1 m. Most of the adventitious roots grow from the branches, forming a rather extended root system with increased space for water and nutrient uptake in dry and early environments, and the width of the root system is usually several times the width of the aboveground crown. Aboveground older branches are highly lignified, grayish-yellow, and stout, and newer branches are slender and grayish-green. Centraliasiaticus is a super-arid desert plant found in deserts and desert grassland zones, preferring to grow on loose, gravelly alluvial and floodplain soils. It often forms communities along dry riverbeds and flowlines, and is also found on stony hills and pre-hill floodplain slopes (**Figure 5**) [26].

4.2 Characterization and phylogenetic appreciation of the chloroplast genome of *A. centraliasiaticus*

We sequenced the chloroplast genome of *A. centraliasiaticus* leaves collected from Yueliang Bay Park in Guide County, Hainan Tibetan Autonomous Prefecture, Qinghai Province, China (101°53′47“ E, 36°09′23” N, 2050 m a.s.l.). The chloroplast genome was sequenced. And the assembled chloroplast genome and its detailed annotation were submitted to GenBank with the accession number OP909739. The

Figure 5.
Morphology and habitat map of Asterothamnus centraliasiaticus. *From Yueliang Bay park in Guide County, Hainan Tibetan autonomous prefecture, Qinghai Province, China, photograph by Zheng-sheng Li.*

results indicate that the complete cp genome of *A. centraliasiaticus* is 152,205 bp in length (**Figure 6**) and comprises a pair of inverted repeats (IR) regions of 25,031 bp each, a large single-copy (LSC) region of 83,956 bp and a small single-copy (SSC) region of 18,187 bp. The GC content of *A. centraliasiaticus* is 37.32%. A total of 130 genes were successfully annotated containing 85 protein-coding genes, 37 transfer RNA genes, and 8 ribosomal RNA genes. Among these genes, 21 genes have one intron each, and two genes contain two introns.

The complete chloroplast sequences of *A. centraliasiaticus,* and other seventeen species in six genera (eight *Aster* species, two *Artemisia* species, one *Pericallis* species, one *Guizotia* species, two *Ambrosia* species, and three *Cynara* species) within *Asteraceae*, were used in the phylogenetic analysis. Two *Oryza* species (Oryza punctata and Oryza minuta) of Poaceae were used as outgroups. The phylogenetic analysis indicated that *A. centraliasiaticus* was close to *Aster hypoleucus* and *Aster lavandulifolius.*

The complete chloroplast sequences of *A. centraliasiaticus,* and other seventeen species in six genera (eight *Aster* species, two *Artemisia* species, one *Pericallis* species, one *Guizotia* species, two *Ambrosia* species, and three *Cynara* species) within *Asteraceae* were used in the phylogenetic analysis. Two *Oryza* species

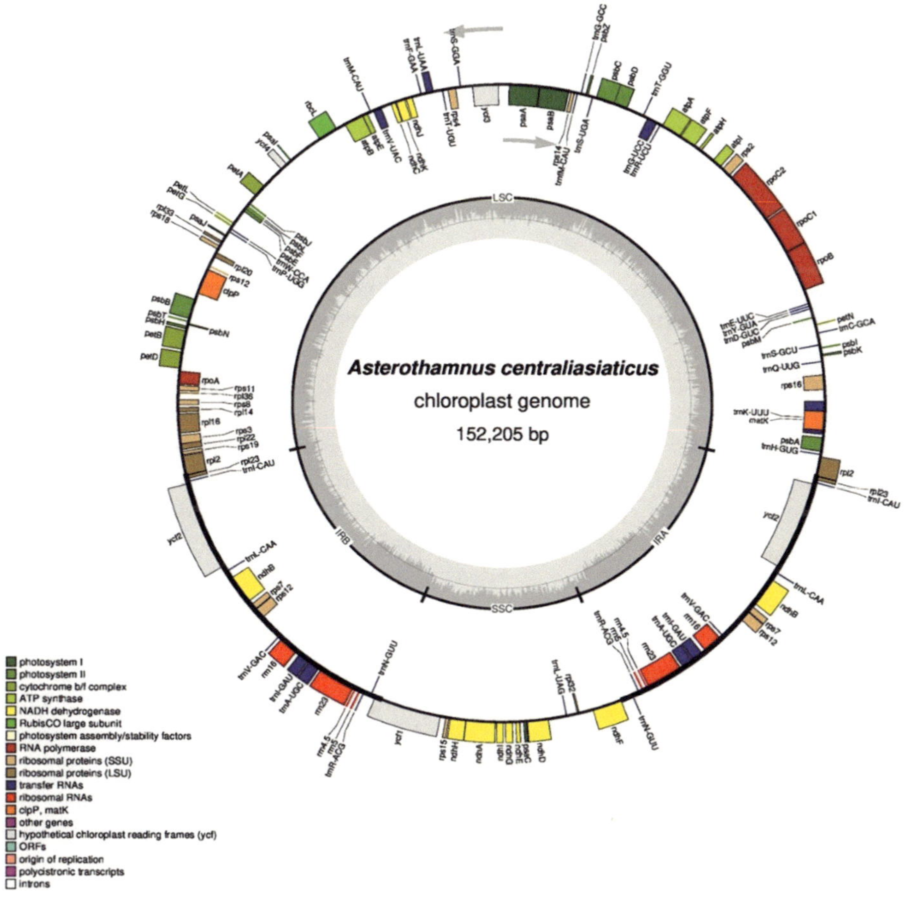

Figure 6.
Chloroplast genome map of Asterothamnus centraliasiaticus.

(Oryza punctata and Oryza minuta) of Poaceae were used as outgroups. The phylogenetic analysis indicated that *A. centraliasiaticus* was close to *Aster hypoleucus* and *Aster lavandulifolius.*

4.3 Discussion and conclusion

Sequencing chloroplast genomes from various plants has provided valuable insights into chloroplast biology, biodiversity conservation, and genetic information that can be harnessed for improving agronomic traits or developing high-value agricultural and biomedical products. In this study, we assembled the chloroplast genome of *A. centraliasiaticus* using Illumina HiSeq2500 sequences. Our findings reveal that the complete cp genome of *A. centraliasiaticus* spans a length of 152,205 bp. Consistent with previous studies, it exhibits a standard quadripartite structure, consisting of a pair of inverted repeat (IR) regions of 25,031 bp each, a large single-copy (LSC) region of 83,956 bp, and a small single-copy (SSC) region of 18,187 bp. We successfully annotated a total of 130 genes, including 85 protein-coding genes, 37 transfer RNA genes, and 8 ribosomal RNA genes. Moreover, the chloroplast genome sequence has been widely utilized for determining evolutionary relationships among plants. In this study, the maximum-likelihood (ML) phylogenetic analysis based on the complete chloroplast genome data strongly supported the close relationship between *A. centraliasiaticus* and *A. hersileeoides*. These findings provide valuable genetic information for germplasm protection and informed development strategies. Furthermore, our analysis revealed that some species of the genus Aster are more distantly related at the chloroplast genome level compared to *A. centraliasiaticus*. This discovery challenges the current botanical species classification of *A. centraliasiaticus* and offers new insights and theoretical support for future plant taxonomy research.

5. Ecology, structural characteristics, and chloroplast genomes of *Corethrodendron multijugum* (maxim.) species

5.1 Morphological characteristics and distribution information

Corethrodendron multijugum (Maxim.) [29] (**Figure 7**), formerly known as *Hedysarum multijugum* [30, 31], also known as *Hedysarum multijugum f. albiflorum*. This genus *Hedysarum* was recently revised to the genus *Corethrodendron* based on morphological and molecular evidence of several barcoding regions, including plastid and nuclear regions [29]. Now, this species belongs to the *Corethrodendron* of *Fabaceae* in *Fabales.*

C. multijugum is a semishrub or herb, woody only at the base, 0.3–1 m tall. Roots lignified. Stem erect, much branched, slenderly striate, densely white pilose, longitudinally furrowed. Leaves 6–18 cm long; stipules brown scaly, ovate-lanceolate, 2–5 mm long, base connate, apex free, abaxially pilose; leaf rachis grooved, densely gray-white pubescent; leaflets 15–35, elliptic, ovate, or obovate, 5–12 mm long, 3–6 mm wide, apex obtuse or slightly concave, base subrounded or rounded-cuneate, ventrally glabrous, abaxially densely appressed pubescent; The petiole is very short and hairy. Racemes growing in axils of branches, 20–35 cm long, 9–25 flowers growing sparsely; bracts caducous; pedicels 2–3 mm long, pilose; calyx obliquely campanulate, 5–6 mm long, calyx teeth subulate or acute, 3–4 times shorter than calyx tube, lower calyx teeth slightly longer than or twice as long as upper calyx teeth, usually

Figure 7.
Morphology and habitat map of Corethrodendron multijugum. *From Yueliang Bay park in Guide County, Hainan Tibetan autonomous prefecture, Qinghai Province, China (101°43'95" E, 36°04'61" N), photograph by Ying Liu.*

splitting between upper calyx teeth deeper below middle of calyx tube, sometimes splitting deeper between both calyx teeth and upper calyx, outside covered with Corolla 15–19 mm long, purple-red or rose-red, with yellow spots; flagellum obovate, slightly concave at first end, 14–18 mm long, claw short; wing petals narrow, 6–8 mm long, ca. 1 mm wide, claw half as long as limb, auricles nearly as long as petiole; keel petal slightly shorter than flagellum, anterior lower angle bow-shaped curved. Ovary linear, pubescent. Pods flattened, usually l-3-noded; ovate or semiorbicular nodding pods ca. 5 mm long and 4 mm wide, sparsely pubescent, with reticulate and small spines on sides. Fl. Jun-Aug, fr. Jul-Sep [12, 29–32].

The plant is widely distributed in northwestern, northern, and southwestern of China, including Sichuan, Tibet, Xinjiang, Qinghai, Gansu, Ningxia, Shaanxi, Shanxi, Inner Mongolia, Henan, and Hubei, and grows mainly in gravelly floodplains and riverbanks in desert areas around 1800–3800 m above sea level, on sunny slopes, gullies, embankments, gravelly lands, gravelly slopes in grassland areas and in certain deciduous broad-leaved forest areas on dry mountain phi and gravelly river banks in some deciduous broad-leaved forest areas [12, 29–34]. It is distributed in all states and counties of Qinghai Province [12, 29]. It is distributed abroad in Mongolia and Russia. The type specimens were collected from the western part of the Hexi Corridor in Gansu [29–32]. This plant has deep roots, strong cold, and drought tolerance [35–37]. It is not only a widely used in traditional Chinese medicine, but also an excellent feed, soil, and water conservation plant, which has important medicinal and economic values [38–42].

5.2 Characterization and phylogenetic appreciation of the chloroplast genome of *Corethrodendron multijugum*

The complete chloroplast genome was sequenced using the Illumina Hiseq 2500 platform (Illumina, SanDiego, CA) with paired-end reads of 150 bp by Genesky

Biotechnologies Inc., Shanghai, China. And the complete chloroplast genome structure of *C. multijugum* was a circular DNA molecule (**Figure 8**), with a length of 122,994 bp (GenBank accession no. NC069301). Unlike the typical tetrad structure of most angiosperm chloroplast genomes, the *C. multijugum* chloroplast genome does not have the typical tetrad structure consisting of a large single copy (LSC), a small single copy (SSC), and a pair of inverted repeats (IRs), and similarly in *Hedysarum polybotrys* var. *alaschanicum* results [43]. The overall G + C content of the whole genome is 34.5%, and the genome presented a negative AT-skew (−0.002) on the J-strand. The genome contains 110 genes, including 76 protein-coding genes (PCGs), 30 transfer RNA genes (tRNAs), 4 ribosomal RNA unit genes (rRNAs). There are 17 genes containing one intron (s) and one trans-splicing genes rps12, and two genes (ycf2 and ycf3) contain two introns. A total of 65 simple sequence repeat (SSR) markers ranging from mononucleotide to tetranucleotide repeat motif were identified in *C. multijugum* chloroplast genome, and mononucleotide had the most repeats.

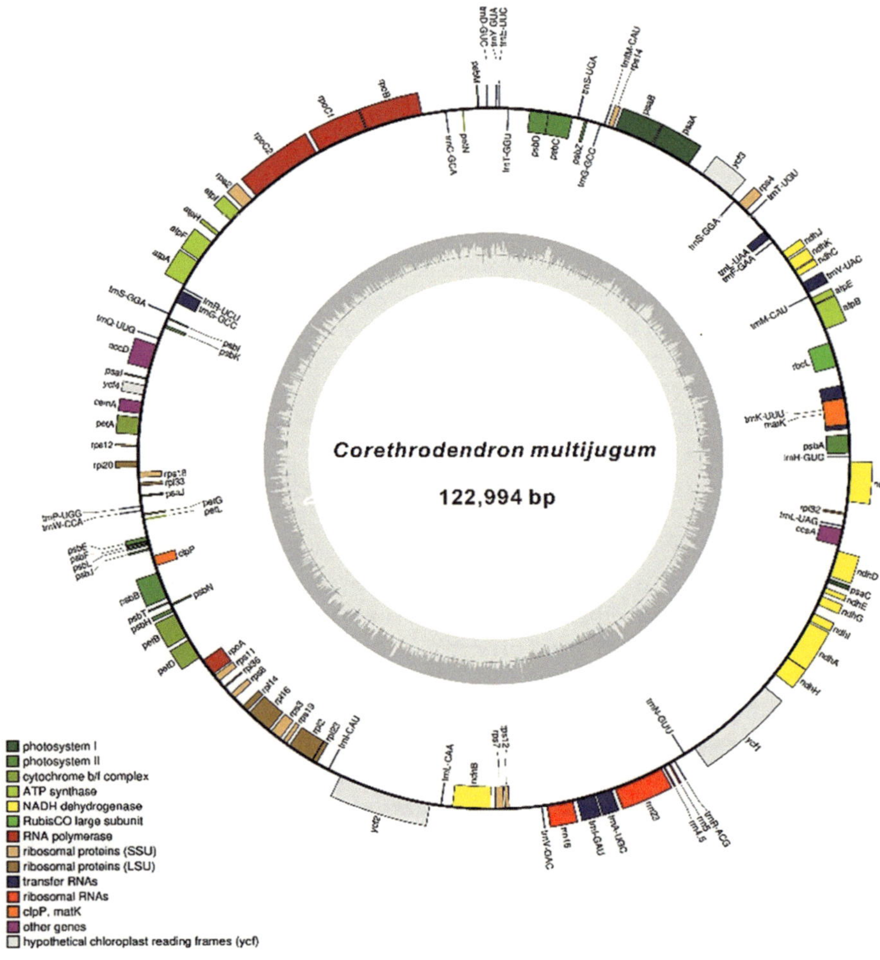

Figure 8.
Chloroplast genome map of Corethrodendron multijugum.

All of the 20 chloroplast genome sequences of Fabaceae were obtained from GenBank and used for phylogenetic analysis, and two chloroplast genome sequences of Rosaceae were used as outgroups. Phylogenetic analysis indicated a strong sister relationship with *C. multijugum* and *Hedysarum petrovii* (**Figure 9**). This study will contribute to a better understanding of the evolutionary pattern of chloroplast genomes in *C. multijugum* and provide more basis for the identification and application of *Corethrodendron* plants.

5.3 Discussion and conclusion

In this study, the chloroplast genome of *C. multijugum* is reported for the first time. We described the sequence structures and annotated genes in the genome. Its sequence length was found to be 122, 994 bp, similar to that of other *Corethrodendron* (*Hedysarum*) species. What's different from other studies is that compared with other previously published chloroplast genomes of *Corethrodendron* (*Hedysarum*) species, the *C. multijugum* chloroplast genome does not have the typical quadripartite structure, and similar genetic structure results have been demonstrated in both genus *Hedysarum* [43] and genus *Astragalus* [44]. What is special about this study is that the phylogenetic tree was reconstructed to confirm the phylogenetic of *C. multijugum* for the first time. The chloroplast genome of *C. multijugum* will contribute to a better understanding of the evolutionary mode of the chloroplast genome and provide more evidence for the identification and application of *Corethrodendron* species.

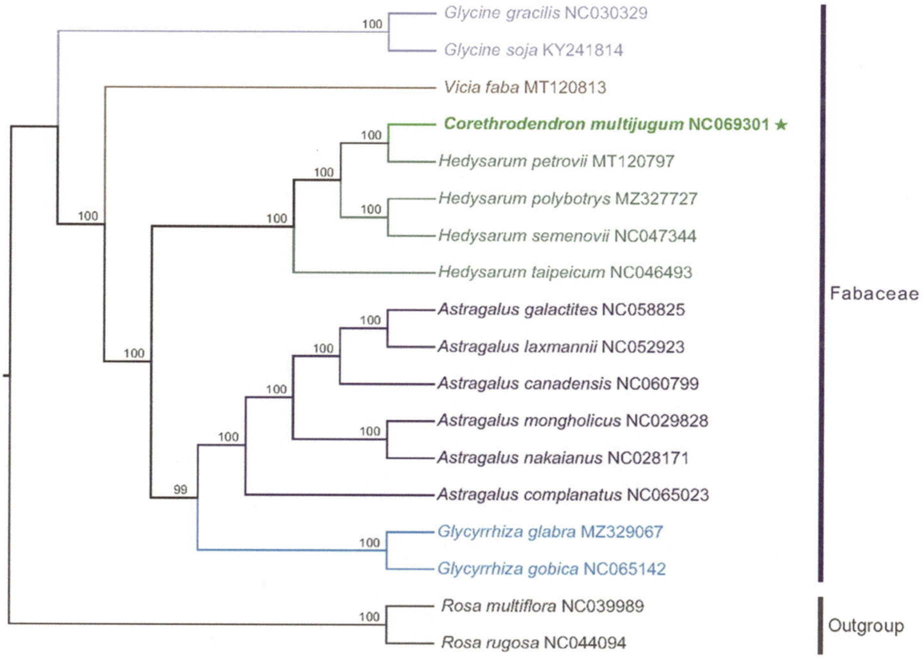

Figure 9.
Chloroplast phylogeny of 16 Fabaceae species based on the complete chloroplast genome sequences. The asterisk represents the assembled plastome sequence in this picture. The clades of species are represented with black lines.

6. Morphological characteristics and chloroplast genome analysis of *Clematis nannophylla* Maxim

6.1 Morphological characteristics and distribution informations

Clematis is belong to widely Ranunculaceae family and distributed worldwide [21, 45]. *Clematis nannophylla* is Small perennial shrub, and has high ornamental, ecological, and medicinal value [22, 45]. *C. nannophylla* is normally found dry or gravelly slopes; 1200–3200 m. Gansu, SW Nei Mongol, Ningxia, Qinghai, Shaanxi (**Figure 10**).

Clematis nannophylla is usually small shrubs, erected 30–100 cm tall. The branches are reddish-brown and ribbed, and the branchlets are densely appressed puberulous and fall off later. Leaves are simple opposite or several fascicled, almost sessile or up to 4 mm long; leaf blade contour subovate, 0.5–1 cm long, 3–8 mm wide, leaf blade pinnately lobed, with lobes 2–3 or 4 in pairs, or lobes subdivided 2–3, lobes or lobules elliptic to broadly obtusely cuneate or lanceolate, 1–4 mm long, with varying 2–3 notched small teeth or entire margin, glabrous or pubescent. Flowers solitary or cyme with 3 flowers; sepals 4, yellow, oblong to obovate, 0.8–1.5 cm long, 5–7 mm wide, pubescent outside, margin densely villous, interior pubescent to nearly glabrous; stamens glabrous, filaments lanceolate, longer than anthers. Achenes oval, about 5 mm long, puberulent. The flowering period is from July to August, and the fruit period is from August to September [22].

6.2 Characterization and phylogenetic appreciation of the chloroplast genome of *C. nannophylla*

The chloroplast genome of *C. nannophylla* was 159,801 bp in length and divided into four distinct regions, such as large single copy region (LSC, 79,526 bp), small single copy region (SSC, 18,185 bp), and a pair of inverted repeat regions (31,045 bp). The genome annotation predicted a total of 133 genes, including 89 protein-coding

Figure 10.
Morphology and habitat map of clematis nannophylla. *From Yueliang Bay park in Guide County, Hainan Tibetan autonomous prefecture, Qinghai Province, China (101°43′95“ E, 36°04´61” N), photograph by Ying Liu.*

genes, 36 tRNA genes, and 8 rRNA genes. Phylogenetic analysis with the reported chloroplast genomes revealed that *C. nannophylla* is nested in Sect. *Fruticella* of family Ranunculaceae and has a close relationship to *C. fruticosa* and *C. songorica* (**Figure 11**).

6.3 Discussion and conclusion

In summary, the complete cp genome sequence of *C. nannophylla* was sequenced and compared with other related species, providing an important reference for the phylogeny of *C. nannophylla*. Although the cp genomes of *C. nannophylla* was identical to each other *Clematis* species in genome structures, gene contents, and GC contents, phylogenetic analysis showed that *C. nannophylla* is closely related to *C. fruticose*, *C. tomentella*, and *C. songarica*, and the well-resolved phylogenetic tree showed monophyletic origin of genus *Clematis* and genus *Aconitum* as its sister genus. The cp genome information in this study provides reference data for molecular marker development, phylogenetic analysis, population study, and cp genome processes, as well as for better exploitation and utilization of *C. nannophylla*. The results can guide more efficient germplasm resource utilization, conservation, and breeding strategy development.

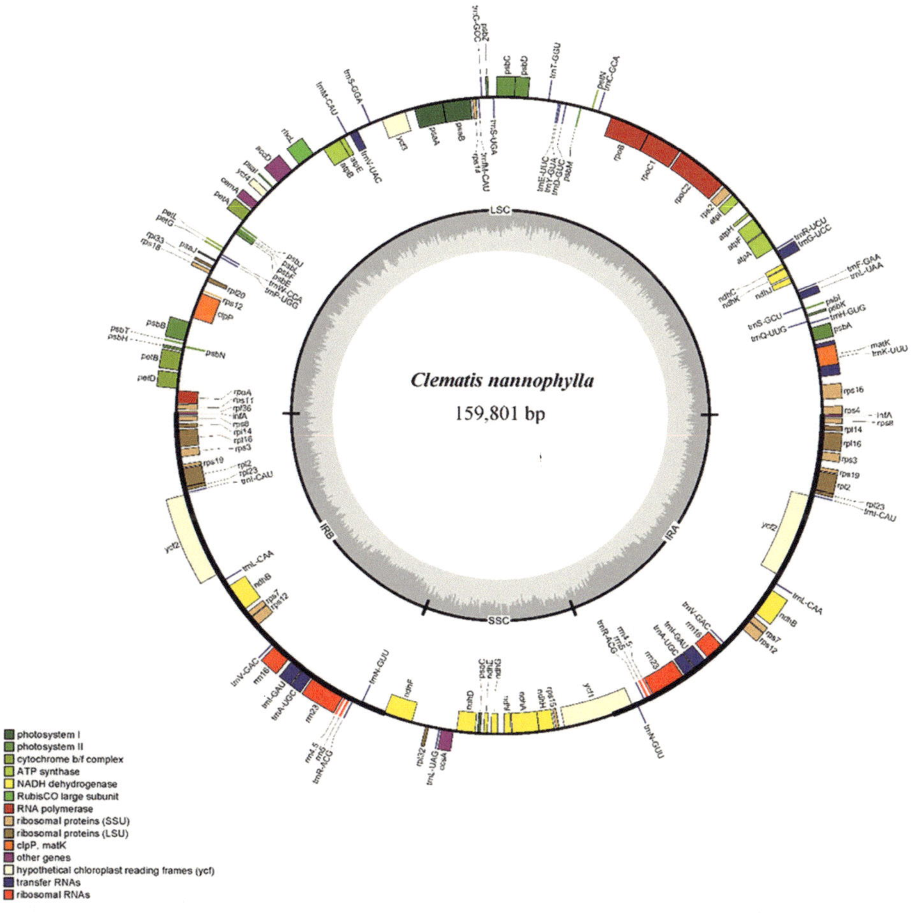

Figure 11.
Chloroplast genome map of clematis nannophylla.

Funding

This work was supported by the Qinghai Science and Technology Department of the senior scientist responsibility system project (2024-SF-101).

Conflict of interest

The authors declare no conflict of interest.

Abbreviations

cp	chloroplast
LSC	large single copy
SSC	small single copy
IR	inverted repeat
SSR	simple sequence repeat
ML	maximum likelihood
RSCU	relative synonymous codon usage
PCGs	protein-coding genes

Author details

Ying Liu*, Jinping Qin, Zhengsheng Li, Lijun Zhang and Xinyou Wang
Qinghai University, Qinghai Academy of Animal and Veterinary Sciences, Qinghai Provincial Key Laboratory of Adaptive Management on Alpine Grassland, Key Laboratory of Superior Forage Germplasm in the Qinghai-Tibetan Plateau, Xining, China

*Address all correspondence to: liuying_yanhong@sina.com

IntechOpen

© 2024 The Author(s). Licensee IntechOpen. This chapter is distributed under the terms of the Creative Commons Attribution License (http://creativecommons.org/licenses/by/3.0), which permits unrestricted use, distribution, and reproduction in any medium, provided the original work is properly cited.

References

[1] Raubeson LA, Jansen RK. Chloroplast genomes of plants, Plant diversity and evolution: genotypic and phenotypic variation in higher plants. 2005. DOI: 10.1079/9780851999043.0045

[2] Daniell H, Lin CS, Yu M, et al. Chloroplast genomes: Diversity, evolution, and applications in genetic engineering. Genome Biology. 2016;**17**:134. DOI: 10.1186/s13059-016-1004-2

[3] Li Y, Zhang L, Wang T, et al. The complete chloroplast genome sequences of three *lilies*: Genome structure, comparative genomic and phylogenetic analyses. Journal of Plant Research. 2022;**135**:723-737. DOI: 10.1007/s10265-022-01417-5

[4] Zong D, Qiao Z, Zhou J, et al. Chloroplast genome sequence of *triploid Toxicodendron vernicifluum* and comparative analyses with other *lacquer* chloroplast genomes. BMC Genomics. 2023;**24**:56. DOI: 10.1186/s12864-023-09154-2

[5] Lin R, Chen YL. Flora of China. In: Phylum Angiosperms, Class Dicotyledons, Compositae, Carduoideae, Trib. Astereae Cass. Vol. 74. Beijing: Science Press; 1985. pp. 133-254

[6] Zhang ZY. Cultivation techniques of *Aster* with high yield. Special Economic Flora and Fauna. 2015;**1**:35-36 (In Chinese)

[7] Ma WJ, Yao GZ, Huang XH, Jia Q, Shen JY, Chang YX, et al. Comparison of the active components of *Aster tataricus* from different regions and related processed products by ultra-high performance liquid chromatography with tandem mass spectrometry. Journal of Separation Science. 2020;**43**(5):865-876. DOI: 10.1002/jssc.201900814

[8] Li L, He L, Su X, Amu H, Li J, Zhang Z. Chemotaxonomy of *Aster* species from the Qinghai-Tibetan plateau based on metabolomics. Phytochemical Analysis. 2021;**33**(1):23-39. DOI: 10.1002/pca.3058

[9] Shen XF, Guo S, Yin YY, Zhang JJ, Yin XM, Liang CL, et al. Complete chloroplast genome sequence and phylogenetic analysis of *Aster tataricus*. Molecules. 2018;**23**:2426. DOI: 10.3390/molecules23102426

[10] An J, Li J, Wang JG, Zhang ZF, Chen C, Zhang H. Authenticating and distinguishing the eight species of traditional Tibetan medicine "Meiduoluomi" by microscopic technique. Microscopy Research and Technique. 2010;**72**(10):727-736. DOI: 10.1002/jemt.20725

[11] Zhang ZF, Luo P, Li J, Yi T, Wang JG, An J, et al. Comparison of the antiinflammatory activities of three medicinal plants known as "meiduoluomi" in Tibetan folk medicine. Yakugaku Zasshi. 2008;**128**(5):805-810. DOI: 10.1248/yakushi.128.805

[12] Northwest Chinese Academy of Sciences and Committee. Qinghai Flora Editorial. Vol. 2. Qinghai flora. Xining, China: Qinghai People's Publishing House; 1997. p. 260

[13] Li WP, Yang FS, Jivkova T, Yin GS. Phylogenetic relationships and generic delimitation of Eurasian *Aster* (Asteraceae: Astereae) inferred from ITS, ETS and trnL-F sequence data. Annals of Botany. 2012;**109**(7):1341-1357. DOI: 10.1093/aob/mcs054

[14] Parmar R, Cattonaro F, Phillips C, Vassiliev S, Morgante M, Rajora OP. Assembly and annotation of red spruce (Picea rubens) chloroplast genome, identification of simple sequence repeats, and phylogenetic analysis in *Picea*. International Journal of Molecular Sciences. 2022;**23**:15243. DOI: 10.3390/ijms232315243

[15] Liu HY, Yu Y, Deng YQ, Li J, Huang ZX, Zhou SD. The chloroplast genome of *Lilium henrici*: Genome structure and comparative analysis. Molecules. 2018;**23**(6):1276. DOI: 10.3390/molecules23061276

[16] Willdenow DCL et al. Enumeratio horti regii botanici berolinensis. Taberna Libraria Scholae Realis. 1809:881

[17] Lin R, Chen YL, Shi Z. Flora of China. Vol. 74. Bei Jing: Science Press; 1985. pp. 112-114

[18] Liu SW, Lu SL, Wu ZL, et al. Qinghai Flora. Vol. 3. Xi Ning: Qinghai People's Publishing House; 1996. pp. 319-322

[19] Ma YQ, F XQ. The Innermogolian Plant Annals. Huhhot: Innermongolian People's Publsihing House; 2000

[20] Jia SX. Forage Plants of China (the First Volume). Beijing: Agricultural Press; 1987. pp. 398-402

[21] Liu D, Qu K, Yuan Y, Zhao Z, et al. Complete sequence and comparative analysis of the mitochondrial genome of the rare and endangered Clematis acerifolia, the first clematis mitogenome to provide new insights into the phylogenetic evolutionary status of the genus. Frontiers in Genetics. 2023;**13**:1050040. DOI: 10.3389/fgene.2022.1050040

[22] Committee of the Flora of China, Chinese Academy of Sciences. Flora of China. In: Ranunculaceae (2), Dicotyledonous Plant Class, Angiosperma Phylum. Vol. 28. Science Press; 1980

[23] Feng J-Y, Wu Y-Z, Wang R-R, Xiao X-F, Wang R-H, Qi Z-C, et al. The complete chloroplast genome of balsam aster (Aster ageratoides Turcz. var. scaberulus (Miq.) Ling., Asteraceae). Mitochondrial DNA Part B: Resources. 2021;**6**(9):2464-2465. DOI: 10.1080/23802359.2021.1955030

[24] Park J, Shim J, Won H, Lee J. Plastid genome of Aster altaicus var. uchiyamae Kitam., an endanger species of Korean asterids. Journal of Species Research. 2017;**6**(1):76-90

[25] Zhang X, Jiang PP, Fan SJ. Characterization of the complete plastome of Aster pekinensis (Asteraceae), a perennial herb. Mitochondrial DNA Part B: Resources. 2021;**6**(3):1064-1065. DOI: 10.1080/23802359.2021.1899081

[26] Lin Y, Chen YL, Shi Z. Flora of China. Vol. 74, Compositaceae [J]. 1985

[27] Jäger EJ, Hanelt P, Davažamc CJF. Zur Flora der Dsungarischen Gobi (Mongolische Volksrepublik). Flora. 1985;**177**(1-2):45-89

[28] Bowman JL et al. Evolutionary changes in floral structure within Lepidium L. (Brassicaceae). International Journal of Plant Sciences. 1999;**160**(5):917-929

[29] Choi BH, Ohashi H. Generic criteria and an infrageneric system for *Hedysarum* and related genera (*Papilionoideae-Leguminosae*). Taxon. 2003;**52**(3):567-576. DOI: 10.2307/3647455

[30] Editorial Board of Flora of China, Chinese Academy of Sciences. *Hedysarum*

multijugum. Vol. 42. Beijing, China: Science Press; 1998. p. 182

[31] Bull. *Hedysarum multijugum* Maxim. Académie Impériale des Sciences Saint-Pétersbourg. 1882;**27**(4):464

[32] Institute of Botany, the Chinese Academy of Sciences. Higher Plants of China. Vol. 2. Beijing, China: Science Press; 1972. p. 436

[33] Tang H. Study on Systematic Taxonomics of the Genus *Hedysarum L.* from China. Northwest A&F University; 2005

[34] Liu P. Phylogeny and Biogeography of the Genus *Hedysarum L.* (Fabaceae). Northwest A&F University; 2017

[35] Cheng J. The excellent herbage of controlling green sand appearance —— *Hedysarum multijugum*. China's Grasslands. 1987;**10**(01):39-42

[36] Pan X, Qiu Q, Li J, et al. Physiological indexes of six plant species from the tibetan plateau under drought stress. Acta Ecologica Sinica. 2014;**34**(13):3558-3567. DOI: 10.5846/stxb201306031288

[37] Gu W, Wang D, Liu X. Effects of drought stress on physiological indexes of resistance in four wild ground cover plants. Northern Horticulture. 2015;**18**:81-83

[38] Wang W, Chen W, Chen H, et al. Studies on chemical constituents of *Hedysarum multijugum*. Journal of Peking University (Health Science). 2001;**33**(03):205-208

[39] Wang W, Chen H, Wang W, et al. Studies on flavoniod constituents of *Hedysarum multijugum*. Acta Pharmaceutica Sinica. 2002;**37**(3):196-198

[40] Wang W, Chen H, Wang W, et al. Studies on chemical constituents and biological activities of *Hedysarum multijugum*. Journal of Pharmaceutical Sciences. 2002;**03**:196-198

[41] Liu X. Excellent wild ecotype ornamental ground cover——*Hedysarum multijugum*. Northern Horticulture. 2007;**1**(07):154-155

[42] Yi L, Wei W, Yuying Z, et al. Chemical analysis of the *Hedysarum multijugum* root by HPLC fingerprinting. Journal of Chinese Pharmaceutical Sciences. 2015;**24**(10):654-659. DOI: 10.5246/jcps.2015.10.083

[43] Cao J-N, Han C-R, Yang Y-C. Characterization of the complete chloroplast genome of *Hedysarum polybotrys* var. *alaschanicum* (Fabaceae) and its phylogeny. Mitochondrial DNA Part B Resources. 2021;**6**(11):3312-3313. DOI: 10.1080/23802359.2021.1994900

[44] Yang J. Characterization of the complete plastid genome of the perennial herb Astragalus complanatus Bunge (Fabales: Fabaceae). Mitochondrial DNA Part B Resources. 2021;**6**(12):3440-3442

[45] Qian R, Ye Y, Hu Q, Ma X, et al. Metabolomic and transcriptomic analyses reveal new insights into the role of metabolites and genes in modulating flower colour of Clematis tientaiensis. Horticulturae. 2023;**9**:14. DOI: 10.3390/horticulturae9010014

Chapter 3

Sequences Evolution and Population Structure of Tunisian Date Palm (*Phoenix dactylifera* L.) Revealed by Chloroplast DNA Markers

Rhouma-Chatti Soumaya, Choulak Sarra and Chatti Khaled

Abstract

Date palm is present among the vital crops of arid and semiarid countries of North Africa and the Middle East. Chloroplast DNA is the best molecule for finding the evolutionary history of plant species. In the present study, cpDNA variation in date palm was estimated using the *trn*L-*trn*F intergenic spacer and *psb*Z-*psb*C region. The high AT values in both molecular markers may clarify the high proportion of transversions observed in this species. The neutrality tests, expansion parameter estimation (mismatch distribution), and haplotype network patterns proposed that demographic expansion had occurred in recent times. Furthermore, the taxa distribution is not related to geographical origins; neighbor-joining trees are clustered independently either from their geographic origin or from the sex of trees, suggesting a common genetic basis between different cultivars. Statistical analysis of chloroplast germplasm provides a means of assessing cytoplasmic gene flow, which occurred in Tunisian *Phoenix dactylifera* L. In fact, N_m was important between Tunisia and Eastern Arabic region ($N_m = 2.57$), which reflects high levels of connectivity between these population pairs. In conclusion, genomic studies prove date palm domestication happened in the Arabian Peninsula and showed an important gene flow with North African palm populations.

Keywords: *Phoenix dactylifera* L., *trn*L-*trn*F spacer, *psb*C-*psb*Z region, population expansion, molecular evolution, gene flow

1. Introduction

Date palms (*Phoenix dactylifera*, Arecaceae) are of major economic and ecological importance to the oasis agriculture of arid and semiarid zones. The past distribution area of this fruit crop covers Mauritania in the west to Pakistan in the east and northwestern India [1]. It is moreover existing in sub-Saharan Africa and has been hosted in California, Peru, Australia, and other countries [2]. The presence of this crop in these areas gives it an undeniable ecological role by limiting the progression of steppe areas and the silting of farmland.

P. dactylifera was brought to Tunisia by Phoenicians before the Roman occupation and has played a vital part in establishing oases. In Tunisia, date palm cultivation covers over 46 thousand ha, with a total number of palm trees of approximately 5.4 million [3]. One hectare of land host an average of 120 date palm trees [4]. Further, the Tunisian date palm germplasm is distinct by a remarkable richness, represented by the high number of cultivars (over 250) [5, 6]. Throughout the plant domestication procedure, breeding actions, selection, migration, and admixture have given growth to cultivated populations distinct from the inherited gene pools [7]. Humans have particularly selected qualities associated with production, fruit quality, and fertility [8].

Understanding the population genetics and domestication history of cultivated species is certainly important for the genetic improvement of crops relying on the conservation and usage of the germplasm [9]. In order to contribute to the varietal improvement of date palms and to offer novel perceptions on the influence of geographic origins and human action on the genetic structure of the date palm, this study investigated the diversity of the species using chloroplast DNA.

The small, sensibly constant size, and conservative evolution of chloroplast DNA (cpDNA) make it an ultimate molecule for finding the evolutionary history of plant species. In angiosperm, the use of chloroplast DNA sequences for intra-specific phylogenetic research is now routine [10]. These investigations have been simplified by the abundant number of whole chloroplast genome sequences that are accessible from an extensive variety of angiosperms and the development of universal PCR primers in conserved coding as well as noncoding regions.

Commonly, in terms of its size, organization, and sequence, cpDNA is the most recognized conservatively evolving genome. It has been used for genetic analysis in plants and provides an accessible and well-characterized source of comparative sequence data. Moreover, identifying the footmark of positive selection is an imperative assignment in evolutionary genetic studies [11]. Otherwise, different modes of selection may result in divergent patterns of the nature and extent of genetic variation. The neutral theory affirms that all observable mutations in populations have little or no effect on an organism's fitness, and their evolutionary dynamics are entirely measured by genetic drift [12, 13]. For such mutations, the evolution continues as equilibrium among the forces of mutation pressure, natural selection, and genetic drift. New molecular techniques based on single and combined sequences data sets have provided a vast understanding of the evolution of flowering plants [14]. Noncoding regions are usually less sensitive to natural selection than coding regions and then may be more beneficial for studying plant evolution.

The chloroplast genome can be classified into three functional categories: (1) protein-coding genes, (2) introns, and (3) intergenic spacers. It has greater phylogenetic potential than nuclear DNA because it is sufficient variable but conserves to be less variable within than between species [15]. The noncoding regions are leading systematic molecular, phylogeographic, and DNA barcoding studies for plants [16, 17]. Chloroplast DNA markers are used for systematic studies of plant species [18–23] and are particularly used to study phylogeny.

In *Paris* genus, Song et al. [24] described eight most variable barcodes, to discriminate between the different species, including psbC-trnS-psbZ region. Moreover, the psbC-trnS intergenic spacer was successfully used for phylogeographic study in *Lolium* species [25]. In 2013, Ballardini et al. confirmed that the psbZ-trnfM (CAU) region could be considered a good basis for the establishment of a DNA barcoding system in *Phoenix*, and is potentially useful for the identification of the female parent in *Phoenix* hybrids.

The date palm chloroplast genome is a typical circular double-stranded DNA molecule, and it shares a common quadri-partite structure: a pair of IRs (27.276 bp) separated by the LSC region (86.198 bp) and the SSC regions (17.712 bp) [26]. An important set of primer pairs improved for PCR amplification and covering sequencing in monocotyledons were optimized, which are distributed throughout the whole chloroplast genome, including exons, introns, and intergenic spacers (IGS).

Thus, in the present study, we undertake to find out the level of cpDNA variation in date palm using the *trn*L-*trn*F intergenic spacer and *psb*Z-*psb*C region and to accurate evolution process, which controls the pattern of polymorphism among the species.

2. Material and methods

2.1 Plant material and DNA extraction

Twenty-four accessions (20 cultivars and four male trees) of Tunisian date palm, listed in **Table 1**, were used in this study. Each cultivar was represented by one tree. Five varieties accommodated in Tunisian plantations ("Ghars Mettig" and "Tantabecht" from Algeria, "Berhi" and "Khadhraoui" from Iraq, and "Abou Meaan" from the United Arabic Emirates) were used in this survey. Young leaves were frozen until their use for DNA purification. Extraction of the total DNA was determined as stated by Dellaporta et al. [27] protocol. DNA concentration and integrity were checked by 0.8% agarose gel electrophoresis according to Sambrook et al. [28].

2.2 Amplification and DNA sequencing

Chloroplastic DNA was sequenced for the *trn*L-*trn*F (intergenic spacer) and *psb*C-*psb*Z (intergenic spacer + gene). Target regions were amplified using universal primers: (Fw1: 5′GGTTCAAGTCCCTCTATCCC3′; Rv1: 5′ATTTGAACTGGTGACACGAG3′) and (Fw2:5′CAACCTTGGCAAGAACG3′; Rv2: 5′TTGACCAACCATCAGRAGA3′) *trnL-trnF* spacer and for *psb*C-*psb*Z region, respectively. Amplification was carried out for 35 cycles, all comprising of a denaturation phase at 95°C for 1 min, annealing at 50°C for 1 min, and an extension step at 72°C for 2 min.

The total volume of PCR reaction was 25 μL, which contained 25 mM of $MgCl_2$, 2 mM of dNTP mix, 1.6 mM of each primer and 1 unit of DNA Taq polymerase, and 20 ng of DNA. Agarose-gel electrophoresis (1.5%) was used to check the PCR products.

The purified PCR products for the *trnL-trnF* and *psb*C-*psb*Z regions were sequenced in both strands according to the automated Sanger method [29] using automated sequencer ABI PRISM™ 310 Genetic Analyzer (Applied Biosystems). The process consists of the selective incorporation of chain-terminating dideoxy-nucleotides by DNA polymerases during in vitro DNA replication. In Sanger sequencing, target DNA is copied multiple times, producing fragments of varying lengths. Fluorescent "chain-terminating" nucleotides mark the ends of the fragments so that the sequence can be determined. It is the most widely used method for detecting single nucleotide variations.

2.3 Sequence analysis

The identity of both sequenced regions was confirmed through a BLASTN search in NCBI database [30]. Nucleotide sequences were aligned by Mega 5.2.2 [31].

Origin	Ecotype	Label	*psb*C-*psb*Z		Combined *Trn*L-*trn*F and *psb*C-*psb*Z	
			Length	%GC	Total length	%GC
Tunisia	Lagou	Lg	953	39.7	1353	37.88
	BesserHelou	Bh	951	40.1	1356	38.14
	Gasbi	Gb	967	40.1	1347	38.21
	Boufeggous	Bf	960	40.1	1360	38.21
	Hamra	Hm	971	40.0	1370	38.14
	Tazerzit Safra	Ts	952	39.8	1335	37.91
	Goundi	Gd	962	39.2	1358	38.18
	Menakher	Mk	967	40.0	1350	37.76
	Ammari	Am	969	39.9	1372	38.11
	Deglet Nour	Dn	971	40.0	1374	38.06
	Kentichi	Kt	964	39.3	1372	38.44
	Oum Laghlez	Ol	965	40.1	1359	38.14
	Arichti	Ar	973	40.1	1373	38.14
	Guelb Jemel	Gj	973	39.8	1355	37.91
	Kharroubi	Kb	967	39.7	1382	38.06
Algeria	Tantabecht	Tb	976	40.2	1357	38.18
	Ghars Mettigue	Gm	952	39.9	1352	37.99
Iraq	Berhi	Br	962	40.1	1345	38.11
	Khadhraoui	Kd	971	39.8	1373	37.99
UEA	Abou Meaan	Ab	969	40.2	1364	38.18
	Male trees					
Tunisia	Borhane2	B2	967	39.9	1368	38.02
	Borhane3	B3	963	39.9	1348	38.06
	CRPh1	C1	966	40.0	1354	38.21
	CRPh5	C5	962	40.0	1362	38.14

Table 1.
Date-palm accessions studied, their origin, accession numbers, and their variation in length, GC and AT contents of the psbC-psbZ region and the combined sequences.

Several genetic parameters were determined with DnaSP program [32]. Haplotype diversity (Hd) [33] and genetic diversity (Pi) [34] were calculated to evaluate genetic diversity. The average of nucleotide differences (k), the minimum number of recombination events (Rm), and the average number of nucleotide differences among cultivars were also detected. By means of selective neutrality tests, we checked the hypothesis of the mutation/drift equilibrium for a supposedly neutral polymorphism. Selection neutrality for the detected mutations was tested by both Tajima's D [35] and Fu and Li's D* and F* methods [36], using the DNAsp program. Demographic parameters were assessed using the distribution of pairwise sequence differences (mismatch distribution) of Rogers and Harpending [37] and site-frequency spectra of Tajima [35].

Moreover, Fu's Fs statistics [38] was used to confirm the assumption of population growth and range expansion as revealed by the mismatch distribution of Rogers and Harpending [37]. In addition, we calculated the Harpending's raggedness index (r) corresponding to an estimate of the fluctuation in the frequency of differences between haplotype pairs [39]. In a complementary way, the R2 statistic [40] was calculated based on the differences between the number of singleton-type mutations and the average of the nucleotide differences. These analyses were executed using coalescent simulations implemented in DnaSP software, with 1000 simulated resampling replicates.

To study the genetic relationships between the studied sequences (haplotypes), we used the reduced median network analysis available in the NETWORK software [41]. The phylogenetic relationships between the studied chloroplast haplotypes were reconstructed using the neighbor-joining (NJ) method [42]. NJ builds a tree from a matrix of pairwise evolutionary distances relating to the set of taxa being studied. Gene flow (*Nm*) was estimated with the mean number of migrants per generation among populations. N_m values were calculated with 1000 data permutations using the software DnaSP v 5.10.01. With reference to the standard for gene flow, we described genetic flow as low for $Nm < 1$, high for $1 < Nm < 4$, and very high for $Nm > 4$ [43]. In gene flow analysis, we consider three different geographic regions of *P. dactylifera* L. Tunisia, Algeria, and Eastern Arabic (Iraq and UAE).

3. Results

3.1 Sequences variation in the psbC-psbZ region

The DNA sequencing of the generated bands has been successfully performed and the blast search allowed confirming the identity of the sequences. DNA sequence varied from 951 bp for "*Besser Helou*" cultivar to 967 bp for "*Gasbi*" cultivar (**Table 1**) with an average of 964.5 pb length. The nucleotide composition frequencies are 0.274 (A), 0.325 (T), 0.204 (C), and 0.196 (G). In addition, the GC content of the amplified sequences varied from 39.7% to 40.7%, and the AT content from 59.3% to 60.3% (**Table 1**).

The polymorphism pattern of the *psb*C-*psb*Z sequence in *P. dactylifera* L. reveals a high level of mutation with 55 polymorphic sites (**Table 2**). We observed 33 singleton variable sites and 22 parsimony-informative ones. Moreover, a low transitional/transversional ratio (ti/tv = 0.68) occurs in the *psbC-psbZ* sequence. In addition, 22 haplotypes were detected among 24 cultivars analyzed, yielding a haplotype diversity of 0.989 (**Table 2**). Sequence variation observed between cultivar groups and nucleotide diversity was estimated (Pi = 0.00862) (**Figure 1a**). The average number of nucleotide differences "k" was estimated to be 8.264. Furthermore, seven regions of conserved DNA were detected using DnaSP program yielding a value of sequence conservation C = 0.913. Mismatch distribution within the species was unimodal for the *psb*C-*psb*Z marker (**Figure 2a**), suggesting demographic expansion had occurred in recent times.

A phylogenetic tree was constructed using *NJ* methods (**Figure 3a**). The evolutionary distances were estimated by the maximum composite likelihood method [44]. Haplotypes of the *psb*C-*psb*Z regions made it possible to make moderately sustained phylogenetic trees (Bootstrap values ≤74%). This dendrogram supported the varieties' organization into two main clusters, which were subdivided into different subclusters. Clearly, the obtained clustering is made independently from the geographical origin of cultivars since the foreign varieties, newly introduced in the

	psbC-psbZ **sequence**	**Combined sequence**
Number of accessions (N)	24	24
Lengths (pb) (Average)	964	1360
GC (%) (Average)	40	37.38
AT(%) (Average)	60	62.66
Number of polymorphic sites (S)	55	69
Parsimony informative sites (Pis)	22	31
Number of haplotypes (H)	22	24
Nucleotide diversity (Pi)	0.00862	0.00861
Haplotype diversity (H_d)	0.989	1
Average of pairwise nucleotide differences (k)	8.264	11.471
Tajima's D	−1.96519 (0.10 > P > 0.05)	−1.74847 (0.10 > P > 0.05)
Fu and Li's D*	−2.345430 (0.10 > P > 0.05)	−1.97901 (0.10 > P > 0.05)
Fu and Li's F*	−2.61261 (0.10 > P > 0.05)	−2.23846 (0.10 > P > 0.05)
Fu's Fs	−12.927 (0.10 > P > 0.05)	−15.474 (0.10 > P > 0.05)
R2 (Ramos-Onsins and Rozas's test)	0.000 P < 0.05	0.000 P < 0.05
Raggedness (r)	0.000 P < 0.05	0.000 P < 0.05

Table 2.
Sequences polymorphism and divergence within date palm cultivars.

Tunisian growing areas, did not remarkably separate from the autochthonous ones. In addition, the male and female trees did not diverge from each other.

In a complementary way, a genetic network based on the *psb*C-*psb*Z sequences was reconstructed (**Figure 3b**). All haplotypes are connected to the most frequent haplotype H1. It represented the ancestor sequence of the other ecotypes, which mutate during evolution. Several putative haplotypes, corresponding to intermediate evolutionary steps, were detected in our network (mv in **Figure 3b**). The haplotype patterns (star-like pattern) reflect a signature of a recent expansion in Tunisian date palm.

3.2 The combined region trnL-trnF and psbC-psbZ spacers

The combined sequence varied from 1347 bp for "*Gasbi*" cultivar to 1382 bp for "*Kharroubi*" cultivar (**Table 1**). Current sequences do not contain insertions or deletions (indels). These sequences revealed 69 polymorphic sites and defined 24 haplotypes. Among the 69 variable sites, 31 were parsimoniously informative and 38 were singletons sites. The means of the haplotype (*Hd*) and nucleotide (*Pi*) diversity (**Figure 1b**) were higher than for *trn*L-*trn*F spacer [16] and the *psb*C-*psb*Z region taken separately. In fact, these values were 1 ± 0.014 and 0.00861 ± 0.001, respectively. The average of pairwise

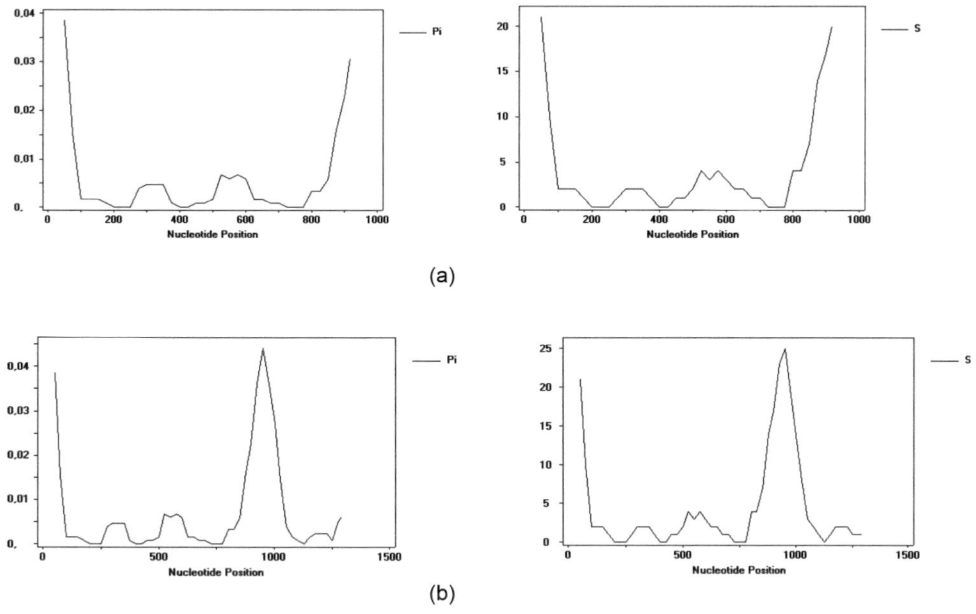

Figure 1.
Variability of nucleotide diversity (Pi) and segregating sites (S) for the psbC-psbZ chloroplastic DNA (a) and the combined sequences (b).

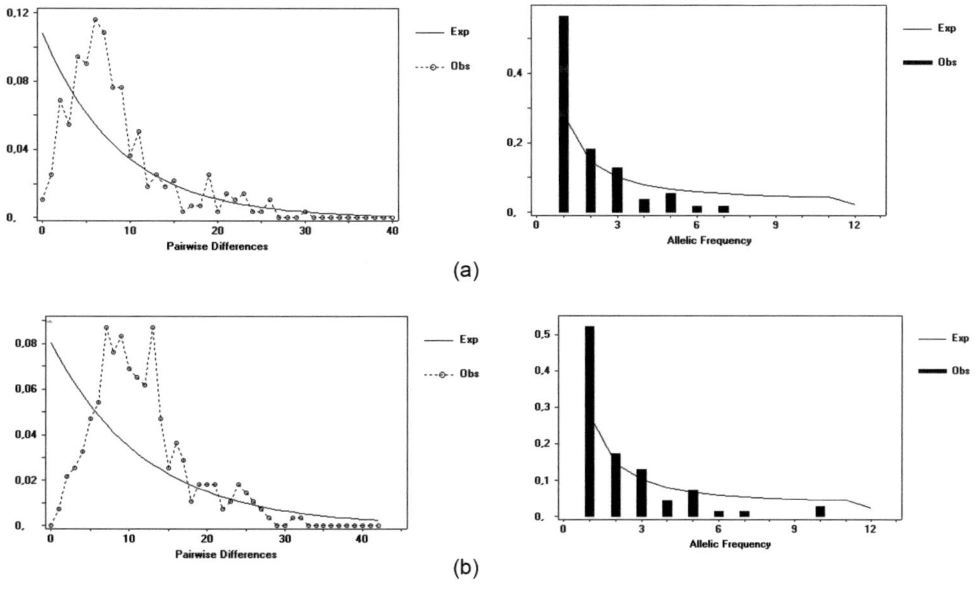

Figure 2.
Mismatch distribution of chloroplastic DNA sequences of the date palm cultivars based on pairwise nucleotide differences in the psbC-psbZ region (a) and the combined trnL-trnF spacer and psbC-psbZ (b). In addition, graphical representation of the site frequency spectrum of chloroplast DNA sequences. Solid lines in the site-frequency spectra indicate the expected distributions under neutrality and at equilibrium.

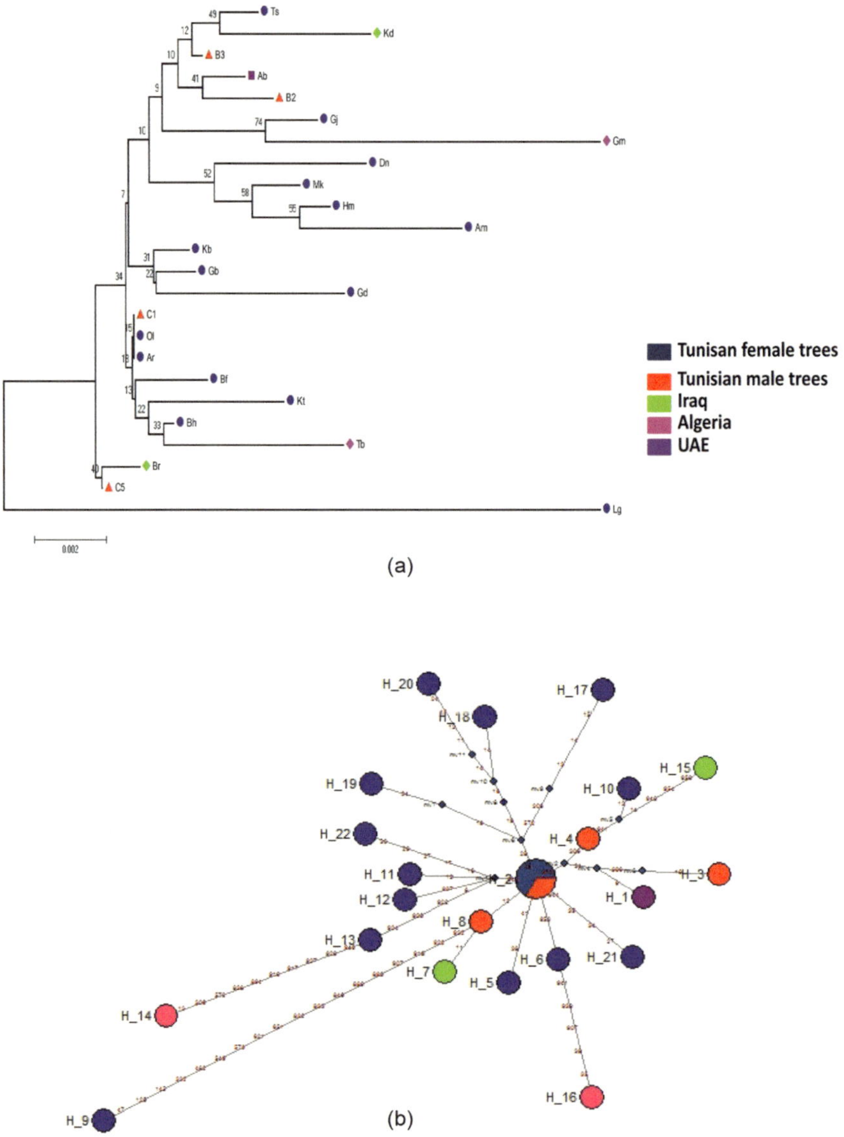

Figure 3.
Summary of phylogenetic analysis within Phoenix dactylifera based on psbC-psbZ sequences: (a) neighbor-joining tree of 24 Tunisian date-palm cultivars. (b) Median-joining network of the haplotypes inferred from the analyzed sequences. Nodes are proportional to haplotype frequencies and branch length is proportional to the number of mutations. The circles represent haplotypes; circle diameters are proportional to the frequencies. Black bands correspond to mutational steps.

nucleotide differences (*k*) was equal to 11.471 (**Table 2**). In addition, the GC content of the combined sequences varied from 37.88% to 38.44% (**Table 1**). The polymorphism pattern in *P. dactylifera* L. revealed an elevated level of mutation points with a low transitional/transversional ratio (ti/tv = 0.63).

Mismatch distributions within the species were generally unimodal for markers, suggesting population expansions (or past selective sweeps) (**Figure 2b**). Numerous

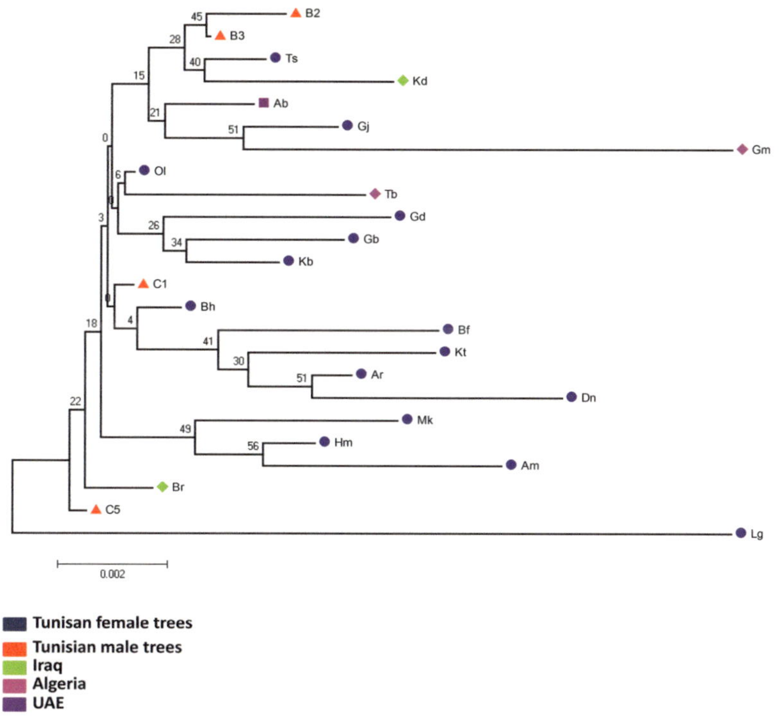

Figure 4.
Neighbor-joining tree based on trnL-trnF spacer and psbC-psbZ gene sequences.

significant negative values of *Fs* and Tajima's *D* in *trn*L-*trn*F [16], *psb*C-*psb*Z spacers, and the combined region (**Table 2**) rule out the hypothesis of a constant population and propose either selective sweeps or past demographic expansion (**Table 2**). In addition, low values of *R2* and Harpending's raggedness index (*r*) will be characteristic of a recent expansion.

For gene flow estimation, N_m was high ($N_m > 1.0$) between Tunisia and Algeria ($N_m = 1.8$) and Tunisia/Eastern Arabic accessions ($N_m = 2.57$) indicating that few differentiations could be established among them. On the other hand, *Nm* was weak between Algeria and Eastern Arabic ($N_m = 0.66$), this reflects low levels of connectivity between these population pairs.

Genetic relationships among date palm cultivars are investigated, also, using the variation observed in the combined (*trn*L-*trn*F and *psb*C-*psb*Z) sequences. As illustrated by **Figure 4**, the phylogeographical patterns were very similar between the *psb*C-*psb*Z gene and the combined sequences. The *NJ* dendrogram supported the variety organization into three monophyletic branches and one cluster. The first branch is composed of "*Lagou*" (Lg), the second one *CRPh5* (C5) Tunisian male tree, while the third is made of Iraquian variety "*Berhi*" (Br).

The remaining cultivars are ranged in the unique cluster, which is divided into four subgroups. In fact, genotypes' clustering is independent of the sex of trees, and it is not structured according to geographical origin. Indeed, with the exception of the foreign cultivar "*Berhi*" (Br) divergence, the other foreign cultivars did not significantly diverge from the Tunisian plantations (**Figure 4**).

4. Discussion

In the present survey, we tested the reliability of the noncoding chloroplast markers to identify date palm cultivars. Therewith, we sought an indication of the level of genetic variation and genetic classification within the date palm cultivars grown in Tunisia.

In this object, we represented the evolution of the *psb*C-*psb*Z gene and the combined sequences (*trn*L-*trn*F spacer/*psb*C-*psb*Z) to reveal polymorphisms in *P. dactylifera* L. The high AT values in both molecular markers may clarify the high proportion of transversions (ti/tv = 0.67 and ti/tv = 0.63) observed in this species. This result corroborates the discovery in angiosperm chloroplast noncoding regions with a ratio ti/tv not exceeding one for any of the examined taxa [45–50]. Base content explains the relatively high proportion of transversions [45].

The use of coalescent theory [51, 52] for the *P. dactylifera* L. sequences has permitted inferences about past and present population size. It has been confirmed that demographic expansions conduct to star-shaped genealogies [53], an excess of rare mutations [54], and a unimodal mismatch distribution [37]. Haplotype network patterns, the neutrality tests, and the expansion parameter estimations (mismatch distribution) proposed that demographic expansion had occurred in recent times as indicated by the excess of mutations type singletons in the *P. dactylifera* sample.

Furthermore, there was no geographical structure in the relationships among the haplotypes; neighbor-joining trees are clustered independently either from their geographic origin or from the sex of trees, indicating a common genetic basis between different cultivars. A low genetic structure is usually related to natural habitat change or to human activities that increase gene flow between populations [55]. Human impact on these regions may be the reason for these findings. Actuality, in Tunisian localities, cultivars are manipulated by farmers after continuous selection, cloning, and exchange of varieties. Similar results have been observed using other molecular markers, where *P. dactylifera* ecotypes clustering made independently from either the tree's sex or their geographic origin in spite of their great distribution [16, 56–60]. This result is contrasted with Zehdi-Azzouzi et al. (2015), where *NJ* classification was clearly coherent with a geographical structuring into two clusters: Eastern pool (Djibouti, Oman, Iraq, the UAE) and the Western pool (North Africa accessions). Also, using three chloroplast regions of rbcL, matK, and trnH-psbA, Iranian date palm cultivars were separated into clades corresponding to their geographical distribution [61].

Statistic analysis of chloroplast germplasm provides a means of assessing cytoplasmic gene flow, which occurred in Tunisian *P. dactylifera* L. The pattern observed confirmed that gene flow is designed to be a significant evolutionary factor in date palm history, allowing a variability of the genetic diversity at different scales (local and foreign). In fact, N_m was important between Tunisia and Eastern Arabic region ($N_m = 2.57$), which reflects high levels of connectivity between these population pairs. Zehdi-Azzouzi et al. [9] proved the movement of gene flows between Eastern and Western origins, mostly from east to west, following a human-mediated diffusion of the species.

In addition, genomic studies prove date palm domestication has happened in the Arabian Peninsula and that the North African population has mixed ancestry with components from Middle Eastern *P. dactylifera* [62]. In this subject, Gros-Balthazard et al. [63] noted that a first domestication event was shown to be improbable, given the very similar form of seeds in cultivars from Africa and the

Middle East. In fact, travelers and pilgrims brought many cultivars of date palms grown in Tunisian oases, particularly from the east [64]. The origins of domestication, the direction of germplasm flows, and the breeding history within the traditional cultivation are sincerely the reason for the observed exchange.

In addition, the mean number of migrants per generation between Tunisia and Algeria (N_m) was important; this could be the result of the relative proximity of geographical sites. In fact, *Deglet Nour* cv. was introduced into Tunisia plantations four centuries ago from western Algeria [64, 65], and became the most valuable cultivar. After this period, the human migrations between Algeria and the south of Tunisia were perturbed by the French occupation of Algeria [64]. Despite that, other works prove that in Algeria, chloroplast diversity presents about 70% of the eastern Arabic chloroplast. In Tunisia and Morocco, this proportion only ranges from 11–42%, although Algerian nuclear diversity is similar to that obtained in Tunisia [66].

In conclusion, environmental change, over-exploitation of water reserves, and an increasing tendency toward *Deglet Nour* cv. Monoculture consists of some of the difficulties of date palm culture in Tunisia. Those are significant issues, which need in-depth research. Considering these, it is decisive to comprehend the genetic diversity and variation among and within accessions, and extensive programs must be prepared, to include collection, estimation, and preservation of plant genetic resources. Determining the gene flow and the demographic expansion of this population should be possible with our two molecular markers, nevertheless, the exploitation of additional nuclear markers clearly provides more accurate results. Actually, with the availability of the date palm nuclear and chloroplast genome sequences, more molecular markers are accessible to explore systematic, cultivar relatedness, and genetic map structure. Furthermore, a webpage regrouping all information about *P. dactylifera* germplasm conservation and utilization has an obligation to be planned in collaboration with all producer countries in order to engender bases for inhibiting its genetic erosion.

Conflict of interest

The authors declare that they have no conflict of interest.

Abbreviations

cpDNA	chloroplast DNA
dNTP	deoxynucleoside triphosphate
Hd	Haplotype diversity
IR	inverted repeats
IGS	Intergenic Spacers
k	average of nucleotide differences
LSC	large-single-copy region
$MgCl_2$	magnesium chloride
Nm	Gene flow
PCR	polymerase chain reaction
Pi	genetic diversity
Rm	minimum number of recombination events
SSC	small-single-copy regions
ti/tv	transition/ transversion

Author details

Rhouma-Chatti Soumaya*, Choulak Sarra and Chatti Khaled
Laboratory of Genetics Biodiversity and Valorisation of Bioresources (LR11ES41), Higher Institute of Biotechnology of Monastir, University of Monastir, Tunisia

*Address all correspondence to: rhoumasoumaya@yahoo.fr

IntechOpen

© 2023 The Author(s). Licensee IntechOpen. This chapter is distributed under the terms of the Creative Commons Attribution License (http://creativecommons.org/licenses/by/3.0), which permits unrestricted use, distribution, and reproduction in any medium, provided the original work is properly cited.

References

[1] Pintaud JC, Ludena B, Aberlenc-Bertossi F, Zehdi S, Gros-Balthazard M, Ivorra S. Biogeography of the date palm (*Phoenix dactylifera* L., Arecaceae): Insights on the origin and on the structure of modern diversity. Acta Horticulturae. 2013;**994**:19-38

[2] Barrow S. A monograph of Phoenix L. (Palmae: Coryphoideae). Kew. Bulletin. 1998;**53**:513-575

[3] Makhlouf-gafsi I, Mokni-ghribi A, Bchir B, Razafindralambo H, Danthine S, Attia H, et al. Foamability and foam stability of male and female date palm sap (*Phoenix dactylifera* L.) during the collection period. Food Biophysics. 2015;**10**(3):360-367

[4] Santoro A, Venturi M, Ben Maachia S, Benyahia F, Corrieri F, Piras F, et al. Agroforestry heritage systems as agrobiodiversity hotspots. The case of the mountain oases of Tunisia. Sustainability. 2020;**12**(10):4054

[5] Rhouma A. Le palmier dattier en Tunisie. Le patrimoine génétique Vol I. Tunis, Tunisie: Arabesques Editions et Créations; 1994

[6] Rhouma A. Le palmier dattier en Tunisie. Rome, Italy: IPGRI; 2005

[7] Doebley John F, Brandon S, Bruce Smith D. The molecular genetics of crop domestication. Cell. 2006:1309-1321

[8] Zohary D, Hopf M, Weiss E. Domestication of Plants in the Old World: The Origin and Spread of Domesticated Plants in Southwest Asia, Europe, and the Mediterranean Basin. 4th ed. Oxford: Oxford University Press; 2012. DOI: 10.1093/acprof:osobl/9780199549061.001.0001

[9] Zehdi-Azouzi S, Cherif E, Moussouni S, Gros-Balthazard M, Abbas Naqvi S, Ludeña B, et al. Genetic structure of the date palm (*Phoenix dactylifera*) in the Old World reveals a strong differentiation between eastern and western populations. Annals of Botany. 2015;**116**(1):101-112

[10] Shaw J, Lickey EB, Beck JT, Farmer SB, Liu W, Miller J, et al. The tortoise and the hare II: Relative utility of 21 noncoding chloroplast DNA sequences for phylogenetic analysis. American Journal of Botany. 2005;**92**(1):142-166

[11] Zeng K, Fu YX, Shi S, Wu CI. Statistical tests for detecting positive selection by utilizing high-frequency variants. Genetics. 2006;**174**:1431-1439

[12] Kimura M. Evolutionary rate at the molecular level. Nature. 1968;**217**:624-626

[13] King JL, Jukes TH. Non-darwinian evolution. Science. 1969;**164**(3881):788-798

[14] Shen P, Wang F, Underhill PA, Franco C, Yang WH, Roxas A, et al. Population genetic implications from sequence variation in four Y chromosome genes. Proceedings of the National Academy of Sciences USA. 2000;**97**(13):7354-7359

[15] Filiz E. Phylogeny of some *Solanum* species (*Solanaceae*) based on complete chloroplast genomes (cpDNA) and individual chloroplast genes. Research in Biotechnology. 2012;**3**(6):33-41

[16] Rhouma-Chatti S, Choulak S, Zehdi-Azzouzi S, Chatti K, Said K. Molecular polymorphism and phylogenetic relationships within

Tunisian date palm (*Phoenix dactylifera* L.): Evidence of non-coding *trn*L-*trn*F regions of chloroplast DNAs. Scientia Horticulturae. 2014;**170**:32-38

[17] Shew J, Lickfy BE, Schilling EE. Comparison of whole chloroplast genome sequences to choose noncoding regions for phylogenetic studies in angiosperms. American Journal of Botany. 2007;**94**(3):275-288

[18] Al-Qurainy F, Khan S, Al-Hemaid FM, Ali MA, Tarroum M, Ashraf M. Assessing molecular signature for some potential date (*Phoenix dactylifera* L.) cultivars from Saudi Arabia, based on chloroplast DNA sequences rpoB and psbA-trnH. International Journal of Molecular Sciences. 2011;**12**:6871-6880

[19] Ahmed A, Enan RM. DNA barcoding based on plastid matK and RNA polymerase for assessing the genetic identity of date (*Phoenix dactylifera* L.) cultivars. Genetics and Molecular Research. 2014;**13**(2):3527-3536

[20] Ahmed A, Enan RM. Cultivar-level phylogeny using chloroplast DNA barcode psbK-psbI spacers for identification of Emirati date palm (*Phoenix dactylifera* L.) varieties. Genetics and Molecular Research. 2016;**15**(3)

[21] Chase MW, Salamin N, Wilkinson M, Dunwell JM. Land plants and DNA barcodes: Short-term and long-term goals. Philosophical Transactions of the Royal Society B: Biological Sciences. 2005;**360**:1889-1895

[22] Chen S, Yao H, Han J, Liu C. Validation of the ITS2 region as a novel DNA barcode for identifying medicinal plant species. PLoS One. 2010;**5**:e8613

[23] Hagar-A H, Gadalla EG, Haggag-S Z, Abu El-Maaty S, Mona HH. Identification of some cultivars of Egyptian date palm (*Phoenix dactylifera* L.) using DNA barcoding. Plant Archives. 2020;**20**:1807-1181

[24] Song Y, Wang S, Ding Y, Xu J, Li MF, Zhu S, et al. Chloroplast genomic resource of *Paris* for species discrimination. Scientific Reports. 2017;**7**(1):3427

[25] Balfourier F, Imbert C, Charmet G. Evidence for phylogeographic structure in Lolium species related to the spread of agriculture in Europe. A cpDNA study. Theoretical and Applied Genetics. 2000;**101**(1):131-138

[26] Yang M, Zhang X, Liu G, Yin Y, Chen K, Yun Q, et al. The complete chloroplast genome sequence of date palm (*Phoenix dactylifera* L.). PLoS One. 2010;**5**(9):e12762

[27] Dellaporta SL, Wood J, Hicks JB. A plant DNA mini preparation: Version II. Plant Molecular Biology Reporter. 1983;**1**:19-21

[28] Sambrook J, Frithsch EF, Maniatis T. Molecular Cloning: A Laboratory Manual. second ed. Cold Spring: Cold Spring Harbor Laboratory; 1989

[29] Sanger F, Nicklen S, Coulson AR. DNA sequencing with chain-terminating inhibitors. Proceedings of the National Academy of Sciences USA 1977;**74**(12):5463-7. DOI: 10.1073/pnas.74.12.5463

[30] Altschul SF, Thomas LM, Alejandro AS, Jinghui Z, Zheng Z, Webb M, et al. Gapped BLAST and PSI-BLAST a new generation of protein database search programs. Nucleic Acid Research. 1997;**25**:3389-3402

[31] Kumar S, Dudley J, Nei M, Tamura K. MEGA: A biologist-centric software for evolutionary analysis of DNA and protein sequences. Brief Bioinformatics. 2008;**9**:299-306

[32] Librado P, Rozas J. DnaSP v5: A software for comprehensive analysis of DNA polymorphism data. Bioinformatics. 2009;**25**:1451-1452

[33] Nei M, Tajima F. Maximum likelihood estimation of the number of nucleotide substitutions from restriction sites data. Genetics. 1983;**105**:207-217

[34] Jukes TH, Cantor CR. Evolution of protein molecules. In: Munroled HN, editor. Mammalian Protein Metabolism. New York: Academy Press; 1996. pp. 31-132

[35] Tajima F. Statistical method for testing the neutral mutation hypothesis by DNA polymorphism. Genetics. 1989;**123**:585-595

[36] Fu YX, Li WH. Maximum likelihood estimation of population parameters. Genetics. 1993;**133**:693-709

[37] Rogers AR, Harpending H. Population growth makes waves in the distribution of pairwise genetic differences. Molecular Biology and Evolution. 1992;**9**:552-569

[38] Fu YX. Statistical tests of neutrality of mutations against population growth, hitchhiking and background selection. Genetics. 1997;**147**:915-925

[39] Harpending HC. Signature of ancient population growth in a low-resolution mitochondrial DNA mismatch distribution. Human Biology. 1994;**1994**:591-600

[40] Ramos-Onsins SE, Rozas J. Statistical properties of new neutrality tests against population growth. Molecular Biology and Evolution. 2002;**23**(8):1642

[41] Bandelt HJ, Forster P, Rohl A. Median-joining networks for inferring intraspecific phylogenies. Molecular Biology and Evolution. 1999;**16**:37-48

[42] Tamura K, Dudley J, Nei M, Kumar S. MEGA4: Molecular evolutionary genetics analysis (MEGA) software version 4.0. Molecular Biology and Evolution. 2007;**24**:1596-1599

[43] Boivin T, Bouvier JC, Beslay D, Sauphanor B. Variability in diapause propensity within populations of a temperate insect species: Interactions between insecticide resistance genes and photoperiodism. Biological Journal of the Linnean Society. 2004;**83**(3):341-351

[44] Tamura K, Nei M, Kumar S. Prospects for inferring very large phylogenies by using the neighbor-joining method. Proceedings of the National Academy of Sciences USA. 2004;**101**:11030-11035

[45] Bakker FT, Culham A, Gomez Martinez R, Carvalho J, Compton J, Dawtrey R, et al. Patterns of nucleotide substitution in angiosperm cpDNA*trn*L (UAA)-*trn*F (GAA) regions. Molecular Biology and Evolution. 2000;**17**:1146-1155

[46] Choulak S, Marzouk Z, Chatti K, Rhouma-Chatti S, Ouni R, Chatti N. Genetic diversity, phylogeography and population gene flow of Tunisian *Pistacia vera* L. Turkish Journal of Botany. 2019;**43**(6):737-748

[47] Choulak S, Rhouma-Chatti S, Marzouk Z, Said K, Chatti N, Chatti K. Chloroplast DNA analysis of Tunisian pistachio (*Pistacia vera* L.): Sequence variations of the intron *trn*L (UAA). Scientia Horticulturae. 2015;**191**:57-64

[48] Hoot SB, Douglas AW. Phylogeny of the Proteaceae based on *atp*B and *atp*B-*rbc*L intergenic spacer region sequences. Australian Systematic Botany. 1998;**11**:301-320

[49] Manen JF, Natali A. Comparison of the evolution of ribulose-1,5-biphosphate carboxylase (*rbc*L) and *atp*B-*rbc*L non-coding spacer sequences in a recent plant group, the tribuRubieae (Rubiaceae). Journal of Molecular Evolution. 1995;**41**:920-927

[50] Morton BR, Clegg MT. Neighboring base composition in strongly correlated with base substitution bias in a region of the chloroplast genome. Journal of Molecular Evolution. 1995;**41**:597-603

[51] Kingman JFC. The coalescent. Stochastic Processes and their Applications. 1982a;**13**:235-248

[52] Kingman JFC. On the genealogy of large populations. Journal of Applied Probability. 1982b;**19**:27-43

[53] Slatkin M, Hudson RR. Pairwise comparisons of mitochondrial DNA sequences in stable and exponentially growing populations. Genetics. 1991;**129**:555-562

[54] Harpending H, Rogers A. Genetic perspectives on human origins and differentiation. Proceedings of the National Academy of Sciences USA. 2000;**96**:10597-10602

[55] Zhang LJ, Cai WZ, Luo JY, Zhang S, Wang CY, Lv LM, et al. Phylogeographic patterns of *Lyguspratensis* (Hemiptera: Miridae): Evidence for weak genetic structure and recent expansion in Northwest China. PLoS One. 2017;**12**(4):e0174712

[56] Rhouma S, Dakhlaoui-Dkhil S, Ould Mohamed Salem A, Zehdi-Azouzi S, Rhouma A, Marrakchi M, et al. Genetic diversity and phylogenic relation-ships in date palms (*Phoenix dactylifera*L.) as assessed by Random Amplified Microsatellite Polymorphism markers (RAMPOs). Scientia Horticulturae-Amsterdam. 2008;**117**:53-57

[57] Rhouma S, Zehdi S, Ould Mohamed Salem A, Rhouma A, Marrakchi M, Trifi M. Genetic diversity in ecotypes of Tunisian date palm (*Phoenix dactylifera* L.) assessed by AFLP markers. Journal of Horticultural Science and Biotechnology. 2007;**82**(6):929-933

[58] Sakka H, Baraket G, Dakhlaoui Dkhil S, Zehdi-Azzouzi S, Salhi-Hannachi A. Chloroplast DNA analysis in Tunisian date-palm cultivars (*Phoenix dactylifera* L.): Sequence variations and molecular evolution of *trn*L (UAA) intron and *trn*L (UAA) *trn*F (GAA) intergenic spacer. Scientia Horticulturae. 2013;**164**:256-269

[59] Sakka H, Zehdi S, Ould Mohamed SA, Rhouma A, Marrakchi M, Trifi M. Genetic polymorphism of plastid DNA in Tunisian date-palm germplasm (Phoenix dactylifera L.) detected with PCR-RFLP. Genetic Resources and Crop Evolution. 2004;**51**(5):479-487

[60] Zehdi S, Cherif E, Rhouma-Chatti S, Santoy S, Salhi Hannachi A, Pintaud JC. Molecular polymorphism and genetic relationships in date palm (*Phoenix dactylifera* L.): The utility of nuclear microsatellite markers. Scientia Horticulturae- Amsterdam. 2012;**148**:255-263

[61] Saboori S, Noormohammadi Z, Sheidai M, Marashi S. Date palm (*Phoenix dactylifera* L.) cultivar relationships based on chloroplast genotyping. Iranian Journal of Science and Technology, Transactions A: Science. 2021;**45**(3):833-840

[62] Flowers JM, Hazzouri KM, Gros-Balthazard M, Mo Z, Koutroumpa K,

Perrakis A, et al. Cross-species hybridization and the origin of North African date palms. Proceedings of the National Academy of Sciences. 2019;**116**(5):1651-1658

[63] Gros-Balthazard M, Newton C, Ivorra S, Pierre MH, Pintaud JC, Terral JF. The domestication syndrome in *Phoenix dactylifera* seeds: Toward the identification of wild date palm populations. PLoS One. 2016;**11**:e0152394

[64] Hamza H, Jemni M, Benabderrahim MA, Mrabet A, Touil S, Othmani A, et al. Date palm status and perspective in Tunisia. In Date Palm Genetic Resources and Utilization. 2015;**2015**:193-221

[65] Kearney TH. Date Varieties and Date Culture in Tunisia. Washington: USDA Bureau of Plant Industry Bulletin; 1906

[66] Moussouni S, Pintaud JC, Vigouroux Y, Bouguedoura N. Diversity of Algerian oases date palm (*Phoenix dactylifera* L., Arecaceae): Heterozygote excess and cryptic structure suggest farmer management had a major impact on diversity. PLoSONE. 2017;**12**(4):e0175232

Chapter 4

Chloroplasts: The Future of Large-Scale Protein Production

Brenda Julian Chávez, Stephanie Solano Ornelas, Quintín Rascón Cruz, Carmen Daniela González Barriga, Sigifredo Arévalo Gallegos, Blanca Flor Iglesias Figueroa, Luis Ignacio Siañez Estrada, Tania Siqueiros Cendón, Sugey Ramona Sinagawa García and Edward Alexander Espinoza Sánchez

Abstract

Chloroplast engineering has matured considerably in recent years. It is emerging as a promising tool to address the challenges related to food security, drug production, and sustainable energy posed by an ever-growing world population. Chloroplasts have proven their potential by efficiently expressing transgenes, encapsulating recombinant proteins, and protecting them from cellular machinery, making it possible to obtain highly functional proteins. This quality has also been exploited by interfering RNA technology. In addition to the practical attributes offered by chloroplast transformation, such as the elimination of position effects, polycistronic expression, and massive protein production, the technique represents an advance in biosafety terms; however, even if its great biotechnological potential, crops that have efficiently transformed are still a proof of concept. Despite efforts, other essential crops have remained recalcitrant to chloroplast transformation, which has limited their expansion. In this chapter, we address the most recent advances in this area and the challenges that must be solved to extend the transformation to other crops and become the de facto tool in plant biotechnology.

Keywords: chloroplast, engineering, agronomical traits, biofuels, biopharmaceuticals, mass protein production

1. Introduction

It has recently been projected that the world population will reach 9.8 billion people in the next three decades, considering that plants provide about 80% of the food that is consumed and that traditional agriculture is inefficient [1–3], meeting their food and public health needs constitutes a challenge. Therefore, seeking to meet the requirements of this growth, biotechnology has used plants as platforms for the production of proteins, and this has been exploited by different industries such as energetic, pharmaceutical, and agricultural, which have already obtained fortified foods, vaccines, antibodies, plants that exhibit resistance to pests and diseases as well as plants with low content of cell wall components [4].

Modified plants have usually been developed by inserting genes directly into the nuclear genome because of the facilities this offers, such as the possibility of using different transformation methods, high transformation frequencies, and the fact that different plant species can be transformed using a single vector type. However, the use of chloroplast transformation seems to be increasing because, although it has lower levels of transgene integration and there are difficulties in obtaining homoplasmic plants [5], it allows the containment of transgenes, which can even be expressed in a polycistronic way, it avoids position effects, and above all, increase the production levels of recombinant proteins. In addition, an advantage that can be exploited is that chloroplasts can function as biocapsules, accumulating proteins of therapeutic interest, and these can pass through the digestive tract without disturbance [6, 7].

Although chloroplast transformation has been implemented in a broader set of plant species, it has not yet been satisfactorily implemented in important crops, such as cereals, considering that these provide about 50% of the proteins in the human diet [8], limits the expansion of this technology. However, although chloroplast transformation still has challenges, it offers an opportunity to lower the production costs of recombinant proteins without biosafety concerns. In this chapter, we explore the advances in the genetic engineering of chloroplasts to produce proteins of industrial interest and the challenges that must be overcome to establish this technology as the preferred tool in the expression of recombinant proteins.

2. Genetics of the chloroplast

Chloroplasts are small organelles primarily associated with their role in the photosynthetic process, which is possible because they can obtain electrons from water [9, 10]. However, other major metabolic processes also occurred within them, such as the biosynthesis of phytohormones, vitamins, pigments, starch, or phenols [11, 12]. Also, the biochemical processes that occur in the chloroplast largely determine the plant's adaptation to its environment [13, 14].

Even though the chloroplasts are semi-autonomous, the control of chloroplast biogenesis and the chloroplast metabolic processes is mainly under nuclear control because, from the ~3000 proteins that there are in the chloroplast, 97% of them are encoded by the nuclear genome and imported via the Toc/Tic machinery [15–17]. Nevertheless, some critical proteins are produced in the chloroplast. There are involved in the transcription, replication, and translation, as well as the proteins that make up the complexes of NADPH plastoquinone oxidoreductase, cytochrome b_6f, and the ATP-synthase subunit, as well as the photosystems I and II, envelope membrane proteins, ß-subunit of acetyl-CoA-carboxylase, and cytochrome C biogenesis [16, 18].

The proteins produced in the chloroplast are encoded by ~120 genes, which are arranged in a circular double-strand genome of quadripartite structure [16, 19, 20] of which, depending on the species and the tissue age, there are multiple copies [21], *e.g.*, 22 copies in potato leaves and 900 copies in wheat leaves. However, the reason for this is not yet known [22, 23].

It has been considered that most chloroplast genomes are highly conserved in terms of their organization and content. However, the length genome varies from

~150 to 220 kb and always appears to be associated with the contraction or expansion of the Inverted Repeat regions [11, 16, 24], which are essential to gene conservation, replication initiation, stabilization, and the evolution of the chloroplast genome [21, 25, 26]. Nevertheless, this is still under discussion because there are species of angiosperms (pea and alfalfa) and gymnosperms (pine) that lack these regions and species such as *Euglena gracilis* that have three sequences but, like repeated sequences clustered in tandem [27–29].

Chloroplast genomes are organized into nucleoids whose number, size, composition, and structural organization are varied, even between adjacent nucleoids [16]. These nucleoids are attached to the thylakoid membrane through different proteins such as MFP1, TCP34, and pTAC16, and seem to be involved in the regulation of gene expression through their association with different SWIB-domain proteins, CND41, sulfite reductase, and WHIRLY proteins [30–33]. In addition, there is strong evidence that nucleoids serve as platforms for forming ribosomes [34, 35]. Therefore, it is currently accepted that nucleoids participate in replication, transcription, translation, post-translational regulation of gene expression, and repair of chloroplast genomes.

Although the number of nucleoids and genomes per nucleoid is variable, it is considered that a nucleoid comprises 10 to 20 copies of plastid DNA. The large number of chloroplast genomes with the possibility of polycistronic expression has made chloroplast genetic engineering rapidly growing [36, 37].

3. Incorporation of new traits in plants through chloroplast transformation

The growing demand for agricultural products has influenced the search for alternatives that maximize their production, including developing modified plants [38]. These crops have allowed an increase in agricultural production, which has also translated into higher profits for farmers [39, 40], to such a degree that by 2018, the commercialization/planting of modified plants had already been adopted by 71 countries [41], which is a rapid growth rate given that it was introduced to the market in 1995, just 23 years prior. The incorporation of new characteristics to plants has maintained an interest in modified crops, so much so that to date, 88% of cotton and 82% of corn grown in the United States is genetically modified [42], and this trend continues, in March 2023 Brazil approved the wheat-HB4 (*Triticum aestivum*) crop that exhibits the *Hahb-4* transcription factor from *Helianthus annuus* that confers drought stress tolerance, as reported by the International Service for the Acquisition of Agri-biotech Applications (event: IND-ØØ412–7) [43].

The development of the first modified plants was achieved by nuclear transformation with the intention of conferring resistance to biotic and abiotic stresses [44–46], and although this method of improvement has been maintained the leadership of plant genetic engineering, to date, we are still grappling with problems such as gene silencing, random transgene integration, inappropriate gene expression regulation, genomic instability, interference with other genes, and selection issues. These problems can affect the expression and function of the transgene, as well as its stability and heritability. Also, there is an ecological risk of the crop-to-crop gene flow [47–49].

To reduce the problems associated with nuclear transformation, biotechnology has gone deeper into the genetic engineering of chloroplasts, which considerably improves the expression levels of recombinant proteins and decreases the unwanted effects associated with modified crops.

The genetic engineering of chloroplasts seems to be growing rapidly, which is possible thanks to the standardization of DNA delivery protocols, the understanding of the mechanisms that govern transformation efficiency, as well as the increase in sequenced chloroplast genomes, which have gone from 6768 in 2021 to 10,712 reported in the RefSeq database from the National Center for Biotechnology Information (NCBI) today, and of which 1072 of them have been reported only so far in 2023.

Although establishing chloroplast transformation in different crops has been challenging, it has already been successfully established in different species (**Table 1**), and currently, been developed plants that express enzymes of industrial value, antigenic

		Family	Species	Ref.
Plants	*Dicots*	Solanaceae	*Nicotiana sylvestris, Nicotiana tabacum* var. Petit Havana, and *Nicotiana benthamiana*	[50–53]
			Petunia hybrida var. Pink Wave	[54]
			Solanum melongena	[55]
			Solanum lycopersicum	[56]
			Capsicum annuum var. G4	[57]
			Solanum tuberosum cv. Desirée and line 1607	[58, 59]
		Brassicaceae	*Brassica napus* cv. FY-4	[60]
			Brassica oleracea var. capitata	[61]
			Brassica oleracea var. botrytis	[62]
			Lesquerella fendleri	[63]
			Brassica napus	[64]
			Arabidopsis thaliana	[65]
		Fabaceae	*Medicago sativa* cv. Longmu 803	[66]
			Glycine max	[67]
		Apiaceae	*Daucus carota* cv. Half long	[68]
		Malvaceae	*Gossypium hirsutum* cv. Coker310FR	[69]
		Asteraceae	*Artemisia annua*	[70]
			Lactuca sativa cv. Verônica, cv Flora and cv. Cisco	[71–73]
		Plantaginaceae	*Scoparia dulcis*	[74]
		Cucurbitaceae	*Momordica charantia*	[75]
		Amaranthaceae	*Beta vulgaris*	[76]
		Salicaceae	*Populus alba*	[77, 78]
	Monocots	Poaceae	*Saccharum officinarum*	[79]
			Oryza sativa var. Japonica line 19 and Hwa-Chung	[80, 81]
Liverworts		Marchantiaceae	*Marchantia polymorpha*	[82]
Moss		Funariaceae	*Physcomitrella patens*	[83]
Algae	*Microalgae*	Isochrysidaceae	*Tisochrysis lutea*	[84]

DOI: http://dx.doi.org/10.5772/intechopen.111829

	Family	Species	Ref.
	Monodopsidaceae	*Nannochloropsis oceanica*	[85]
	Phaeodactylaceae	*Phaeodactylum tricornutum*	[86]
Red algae	Cyanidiaceae	*Cyanidioschizon merolae*	[87]
	Bangiaceae	*Pyropia yezoensis*	[88]
	Porphyridiophyceae	*Porphyridium* sp. UTEX 637	[89]
Green algae	Dunaliellaceae	*Dunaliella tertiolecta*	[90]
	Haematococcaceae	*Haematococcus pluvialis*	[91]
	Euglenaceae	*Euglena gracilis*	[92]
	Chlamydomonadaceae	*Chlamydomonas reindhartii*	[93]

Table 1.
Species transformed by stable chloroplast transformation.

proteins for vaccine production, proteins of pharmaceutical interest, and proteins for the production of biofuels and biomaterials [36, 94–97], as well as proteins that confer resistance to pests and diseases [77, 98, 99].

3.1 Chloroplast transformation to confer resistance to pests

Annually, about 40% of agricultural products are lost before harvest due to attacks by insects, weeds, and plant pathogens, increasing by 20% after harvest [100]. Therefore, using pesticides has been one of the alternatives that have allowed controlling these losses, increasing their use in cropland from 1.55 kg Ha^{-1} in 1990 to 2.69 kg Ha^{-1} in 2019 [101].

The use of pesticides has contributed significantly to agricultural production. However, the emergence of pests resistant to these chemicals (>17,000 cases of resistance amongst 612 species globally by 2020) [102] has resulted in a greater reliance on higher doses and the introduction of new pesticides, which is a worrying trend since it can lead to environmental contamination, and more significant risks for human health; by the way, each year, millions of people around the world experience unintentional acute poisoning by pesticides (~385 million reported cases), of which 11,000 cases end in death [103]. Although this figure is already worrying in itself, it must be considered that 50% of pesticides are organophosphates. These are of particular concern since they pass from the roots to the leaves and can be consumed by the public, which increases the risk of poisoning [104–106].

Pursuing sustainable and effective alternatives to protect crops without the negative consequences of excessive pesticide use resulted in the introduction of the first insecticidal gene, Cry1Ab, to plants through nuclear transformation [44]. Since then, numerous studies have demonstrated the efficacy of Cry proteins against various pests, including *Plutella xylostella Sesamia inferens*, *Chilo suppressalis*, *Herpitogramma licarisalis*, *Mycalesis gotama*, *Cnaphalocrocis medinalis*, *Scirpophaga incertulas*, *Naranga anescens*, *Parnara guttata,* and *Elasmopalpus lignosellus* [107–111]. Cry proteins have also been tested to control nematodes, Phthiraptera, Orthoptera, mites, Coleoptera, Lepidoptera, Diptera, Hymenoptera, Hemiptera, and protozoa [112, 113]. The usefulness of these proteins has led to their continued study and exploration, with over 700 insecticidal proteins already reported (http://www.lifesci.sussex.ac.uk/home/Neil_Crickmore/Bt/). However, insects' resistance to Bt crops has been reported (from 3 cases in 2005 to 16 in 2016), affecting crops containing Cry1Ab, Cry1Ac, Cry1A.105, Cry1Fa, Cry2Ab, Cry3Bb, mCry3A, eCry3.1Ab, and Cry34/35Ab [114, 115]. On the other hand, there is a direct observation of resistance by *Diabrotica virgifera virgifera* to the eCry3.1Ab protein in the field before the plants were commercialized [116, 117]; such resistance may increase through cross-resistance.

Although to reduce the risks of insect resistance to Bt crops have been used strategies such as multiple toxins, protein engineering, ultra-high doses, and refugees [118], the existence of other risks of the nuclear transformation, such as ecological risks and collateral effects in the expression of the genes, have motivated the use of chloroplast engineering for the expression of proteins with insecticidal potential, which began with the expression of the Cry1Ac protein in tobacco chloroplast obtaining protection against lepidopteran *Heliothis virescens, Helicoverpa zea* and *Spodoptera exigua* [45].

Other Cry proteins such as Cry9Aa2, Cry1Ia5, Cry2Aa2, Cry1C, and Cry1Ab have also been expressed in chloroplast from soybean, cabbage, and poplar, including protecting the plants against *Phthorimaea operculella, Hyphantria cunea, P. xylostella, Lymantria dispar, Anticarsia gemmatalis* and *Helicoverpa armigera* [77, 119–123], and until today, the insects did not show resistance to Cry proteins compartmentalized in the chloroplast; therefore, the accumulation proteins in the chloroplast, either through their direct expression within it or by redirecting nuclear-expressed proteins to the chloroplast for storage, seems to be a viable option to avoid the development of resistance. Although this last option it as already been analyzed with the expression of Tvip3A*, Cry1Ac, Cry1Ah, and Cry2A proteins [124–127], and is very attractive because it benefits from the facilities offered by nuclear transformation, the compartmentalization of proteins does not eliminate the risks of environmental contamination, nor any other undesired effect caused by gene insertion into the nuclear genome. Therefore it requires a more detailed analysis.

Other proteins with insecticidal activity have been expressed using chloroplasts, such as a chloroperoxidase from *Pseudomonas pyrrocinia* and the *pta* gene from *Pinellia ternata* agglutinin being effective in the control of *Alternaria alternata, Aspergillus flavus, Escherichia coli, Fusarium moniliforme, Pseudomonas syringae*, *Verticillium dahliae, Colletotrichum destructivum*, and *Fusarium verticillioides* [128–130].

Although the Cry proteins with which Bt crops have been nuclearly armed have shown excellent performance in controlling different pests, the expression of lectins and other proteins with insecticidal capacity from *Bacillus toyonensis* and *Lysinibacillus sphaericus* has also been shown to be effective in controlling pests such as *Alphitobius diaperinus, Spodoptera exigua, Cydia pomonella, Anthonomus grandis*, *Aedes aegypti*, and *Myzus persicae* [131, 132]. Furthermore, while the expression of these proteins in chloroplasts has not been thoroughly investigated, it is important to note them because despite the efficacy of Bt crops in controlling Lepidoptera and Coleoptera, control of Hemiptera has not been entirely successful, as many Hemipteran species have now become significant pests of Bt crops [133, 134].

3.1.1 Expression of RNAi for pest control

In recent years, it has been proposed that the expression of interference RNA (RNAi) for pest control in the chloroplast is a promising alternative to protein expression, and although the RNAi has been achieved previously by nuclear transformation using both double-stranded RNA (dsRNA) and long hairpin RNA (hpRNA)

[135–137], the obtained level protection in plants has been insufficient for practical application [138].

One problem in achieving high levels of protection has been the impossibility of accumulating large amounts of unprocessed RNAi in the host because the nuclear-expressed RNAi are processed by the cellular machinery leaving little raw RNAi available for ingestion by the host [139, 140]. Furthermore, when RNAi is ingested, the host's digestive system degrades another part of the ingested RNAi, resulting in a reduced amount of RNAi available for effective interference [141]. Therefore, the chloroplast has been visualized as a viable solution because it offers three crucial advantages 1) there is no RNAi processing machinery in these organelles, 2) they can produce and accumulate large amounts of RNAi [142–144], and 3) the chloroplast acts as a natural bioencapsulation method to protect the RNAi when consumed by insects [145, 146].

There is limited information on RNAi expression in the chloroplast, but the published reports show an area of high potential. For example, recently [138] expressed dsRNA targeting the *MpDhc64C* gene, a newly identified target gene whose silencing causes the lethality of the green peach aphid *Myzus persicae*. Results revealed that transplastomic plants exhibited significant resistance to aphids, which in turn showed reduced survival, decreased fecundity, and decreased weight of survivors, a similar effect to the obtained by Ren, Cao [147] with the use of dsRNA β-*Actin to control of the beetle Henosepilachna vigintioctopunctata in transplastomic potato plants.*

Other RNAi studies in chloroplasts have shown results that go the same way. In 2015, Jin et al. [141] silenced the V-ATPase, chitin synthase (*Chi* gene), and cytochrome P450 monooxygenase (*P450* gene) from *H. armigera* using dsRNA, reducing the weight and growth of larvae. Also, a hpRNA targeting acetylcholinesterase (*ACE* gene) of *H. armigera* and a dsRNA targeting β-actin and Vps32 (*ACT* and *SHR* genes, respectively) from Colorado potato beetle provided substantial protection against *H. armigera* and reduced growth of larvae of Colorado potato beetle in transplastomic tobacco and potato plants [140, 143].

Although the RNAi expression in chloroplast has been successfully established with promising results of accumulation and stability, future studies should be focused on elucidating the ideal length of the RNAi since there are reports of dsRNA with a length of 200 nt more protective than a dsRNA of 60 nt or > 200 nt; also, it has been reported that the cellular machinery can degrade the long RNAi producing siRNA that have a less insecticidal effect [136, 140]. Therefore, the adequate length of the RNAi that must be expressed is still not entirely clear, and this is important because the length and type of RNAi, whether it is hpRNA or dsRNA, dramatically influences the accumulation [138, 142, 144].

Other aspects that should also be considered are the suitable target genes and the implementation of efficient RNAi delivery methods, even though topical RNAi has recently been reported and seems a cost-effective alternative [148]. Currently, micro-injection and overall ingestion are the most used. Moreover, this last, in practice, should be carefully considered because there are pests that do not have direct contact with chloroplasts and considerably impair RNAi efficacy [149].

In 2017, was approved the first variety of transgenic maize Smartstax PRO®, which nuclearly express a dsRNA of the *Snf7* (Sucrose non-fermenting 7) gene from *D. virgifera virgifera* [150–152], whose commercial launch was in 2022, which supports the idea that RNAi technology has great potential, and although there is no product developed from chloroplast transformation on the market yet, RNAi technology may help chloroplast engineering in this process.

3.2 Chloroplast transformation to confer abiotic stress tolerance

Reactive oxygen species (ROS) are forms of oxygen partially reduced and produced during biotic and abiotic stress. Although ROS are commonly associated with stress, they are also produced under cellular respiration and photosynthetic processes [153]. Nevertheless, although ROS are produced in normal metabolic processes, the uncontrolled production of ROS causes an alteration of the correct function of the cells. Therefore, decreasing ROS in plants is still an essential target in biotechnology.

Nuclearly have been expressed genes to promote the tolerance to different abiotic stresses such as cold stress (*CsWRKY46* gene) [154], oxidative stress (*ScVTC2* gene) [155], drought and salt stress tolerance (*ZmSNAC13* and *ZmWRKY86* gene, respectively) [156, 157], metal tolerance (*OsMYB-R1* and *SbMT-2* gene) [158–160], ROS decrease (*CfAPX* gene) [160] and tolerance to waterlogging (*HaOXR2* gene) [161] and have been obtained satisfactory results; nevertheless, trying to improve the results obtained by nuclear expression, genes have been expressed in chloroplasts aimed at increasing the antioxidant pathway, intervening in the glycine betaine (GB) pathway such as *codA* gene from *Arthrobacter globiformis* [98, 162, 163].

Other proteins have also been expressed in the chloroplast, such as flavodoxin (*fld* gene) [164], arabitol dehydrogenase (*ArDH*) [165], *otsB-A* operon (trehalose phosphate synthase/phosphatase) [166], γ-tocopherol methyltransferase (*TMT* gene) [167], betaine aldehyde dehydrogenase (*badh* gene) [68], dehydroascorbate reductase (*DHAR* gene), superoxide dismutase (*MnSOD* gene), glutathione reductase (*gor* gene) [168], glutathione-S-transferase (*GST* gene) [162, 169], homogentisate phytyltransferase (*HPT* gene), conferring tolerance to salt, cold, UV-B radiation, heavy metal, and osmotolerance.

Despite the potential and benefits of plastid transformation in protecting plants, it is challenging to confer abiotic stress resistance due to the involvement of various metabolic pathways. Therefore, providing a robust resistance would require multiple gene expressions. These must be carefully selected since the expression of certain types of genes can cause pleiotropic effects because the encoded proteins could interfere with the structure and function of the thylakoids or decrease the levels of ATP production [170, 171]. Nevertheless, despite the challenges of achieving stress resistance, the population's constant growth requires finding strategies to address this need.

3.3 Expression of hydrolytic enzymes in chloroplasts

Lignocellulose residues are the most abundant raw material and a highly renewable carbon source on earth [172], and due to particularities as its abundance, availability, and sustainability is believed to be a solution to solve fossil fuel shortage [173–175], to such a degree that global ethanol and biodiesel production is projected to rise to 132 billion liters and 50 billion liters, respectively, by 2030 [176]. However, although lignocellulosic compounds are abundant [177, 178], a large part of this renewable energy is beyond our reach; that is, despite being produced each year more than 40 million tons of non-edible plant material, lignocellulose is not the most important feedstock for biofuel production because it is not efficiently processed [176, 179].

The processing problem is caused by the presence of lignin that imbibes both the cellulose as well as hemicellulose and limits its degradation [180–182]. Therefore, physical and chemical methods have been used to increase biomass degradation.

However, apart from the fact that these methods can represent a potential ecological risk, they increase production costs.

In the search for alternatives to improve biomass degradation, the use of multiple enzymes has been recurring, and in this sense, different organisms have been the source of them [183]; nevertheless, the hydrophobic nature of the substrate, the enzymes cost, the concentration of required enzymes, as well as the release of phenolic compounds during the enzymatic reaction such as xylan that can inhibit enzyme activity, have limited their use [184–186], forcing the search for new and more efficient enzymes as well as an efficient method for their accumulation; therefore, the protein expression in the chloroplast it has presented as a promissory strategy to aboard the challenges of the degradation of lignocellulosic biomass.

The expression of enzymes capable of degrading cell wall polymers into chloroplast has already been tested successfully. However, although different genes of pectinases, manganese peroxidase, cutinase, and laccase enzymes from *Fusarium solani*, *Streptomyces thermocarboxydus, Phanerochaete chrysosporium*, *Trichoderma reesei*, and *Pleurotus ostreatus* has been tested with promissory results [187–190], cellulases have been the subject of the most intensive search because they represent 40% of plant biomass [177, 178].

Different cellulase enzymes have already been expressed in the chloroplastic compartment with satisfactory results, *e.g.*, xylanases (*xynA, xyl, xyn2, Xyl10B, xynA, xyn10A,* and *xyn11B* genes) from *Bacillus subtilis, Trichoderma reesei, Clostridium cellulovorans, Thermotoga maritima, Clostridium cellulovorans, and Alicyclobacillus acidocaldarius* [191–194], as well as endo-, exo-glucanases, and β-glucosidases (*cel6A, cel9A, cel6B, Cel6, Cel7, EndoV, CelK1, Cel3, TF6A, Bgl-1, bgl1C, EGPh, celA,* and *celB* genes) from *Trichoderma reesei, Thermomonospora fusca, Pyrococcus horikoshii, Chaetomium globosum, Paenibacillus* sp., *Phanerochaete chrysosporium*, *Aspergillus niger, and Thermotoga neapolitana* [36, 195–200]. Despite the expressed proteins being functional, it should be considered that the microorganisms that possess efficient mechanisms for cellulose degradation have redundant genes often [201]. Therefore, high degradation depends not on a particular protein but on its enzymatic cocktail, which should be considered to improve enzymatic processes.

Regardless of the intense search for cell wall hydrolytic enzymes that have been carried out, there is still a need to identify efficient cellulose enzymes [184] because they are critical factors in paper recycling, cotton processing, juice extraction, detergent production, food industry, animal feed additives and largely determine the price of biofuels [202–204]. For this reason, it is necessary to use strategies that reduce cellulase enzyme production costs as much as possible to make them commercially viable on a larger scale. Further studies will continue to be carried out to express cellulases in chloroplasts.

3.4 Chloroplast engineering for the biopharmaceutical industry

Population growth implies a constant demand for medicines, representing a lucrative opportunity for the pharmaceutical industry, which only in 2022 billed ~1.48 trillion U.S. dollars, an increase of 4.23% concerning 2021. However, most of the world's population cannot access medicines due to unaffordable prices [205–207]. Therefore, products with nutritional and pharmaceutical value are gaining importance to such an extent that there are currently 1775 products with different formulations, biosimilars, and biobetters approved by the Foods and Drugs Administration (FDA) and European Medicines Agency (EMA) for use in humans [208], which can reduce the costs of medical treatments.

Chloroplasts have already begun to contribute to the production of plasma proteins, vaccines, antibodies, and enzymes [97, 209, 210], which is especially relevant since two of the current challenges in treatments with proteins for oral consumption are obtaining high protein concentrations and the possible degradation that they may experience when passing through the digestive system [206]. These challenges could be overcome by the accumulation and bioencapsulation of proteins in chloroplasts [211].

Two decades have passed since the first candidate antigen against a human disease was expressed [212], and since then, other proteins have been expressed, *e.g.,* A27L immunogenic protein (*A27L* gene) [213], ESAT6, Mtb72F [214, 215], *Angiotensin-converting enzyme 2 (ACE2* gene), Angiotensin (1–7) (*Ang-*[*1–7*] gene) [7, 211], Coagulation factor VIII (*F8* gene) [216], E7 Human papillomavirus antigen (*E7* gene) [217], Coagulation Factor IX (*F9* gene) [96], SAG1 Surface antigen (*SAG1* gene) [95], EDIII-1, EDIII-3, EDIII-4 (*ediii-1, ediii-3,* and *ediii-4* genes) [218], KETc1, KETc7, KETc12, GK1, TSOL18/HP6-Tsol [219], Griffithsin protein (*grft* gene) [220], S1D [221], and Human epidermal growth factor (*hEGF* gene) [94] in plants; and recently, was cloned the gene *IL29* in alga to produce human interleukin 29. The proteins expressed in the chloroplastic compartment have been obtained in high concentrations retaining their functionality, showing the potential value of expressing biopharmaceutical proteins in chloroplasts [222].

One of the challenges currently faced by therapeutic proteins is to stimulate their passage through the epithelial mucosa [223]; however, this challenge has been efficiently addressed by expressing the proteins in the chloroplast fused with the cholera toxin B subunit (CTB), and this has already been tested with the expression of exendin-4 (CTB-EX4) and acid alpha-glucosidase (CTB-GAA) [6, 224]. Fusing the proteins to CTB is an important approach as it has been reported that it could help promote oral tolerance; however, other non-toxic fusion proteins have been reported, such as human transferrin, although there are no reports with this fused protein in the chloroplast.

There is a need for biobetters/biosimilars on the market, and expressing therapeutic proteins in the chloroplast is not only feasible but could solve delivery and efficacy issues. To date, no one plant/chloroplast-based vaccine against human diseases is available to the public. On the other hand, although transplastomic plants have been cultivated for just over a decade with the consent of the United States Department of Agriculture Animal and Plant Health Inspection Service USDA-APHIS, they have not been scaled to the commercial level even though these plants do not fit USDA-APHIS regulation 7 CFR part 340 [206].

4. Conclusions

Given that the world population has been projected to increase by just over 20% in 30 years, it becomes evident that it will be necessary to adopt new technologies to face this situation in areas such as agriculture, energy, and pharmaceuticals, which are key sectors to guarantee well-being and sustainability in the future. For almost three decades, chloroplast transformation has shown important qualities that can help us address the challenges, offering greater efficiency and performance in the expression of modified genes and more precision and control in genetic modification accompanied by greater environmental safety and biosafety. While some challenges need to be addressed, such as the optimization of regulatory regions, limited expression of proteins in non-photosynthetic tissues, the understanding of the mechanisms that govern pleiotropic effects, and the current inability to transform cereals efficiently, it is a fact that it can play a fundamental role in the production of products.

Conflict of interest

The authors declare no conflict of interest.

Author details

Brenda Julian Chávez[1], Stephanie Solano Ornelas[1], Quintín Rascón Cruz[1], Carmen Daniela González Barriga[2], Sigifredo Arévalo Gallegos[1], Blanca Flor Iglesias Figueroa[1], Luis Ignacio Siañez Estrada[1], Tania Siqueiros Cendón[1], Sugey Ramona Sinagawa García and Edward Alexander Espinoza Sánchez[1*]

1 Biotechnology Laboratory I, Faculty of Chemical Sciences, Autonomous University of Chihuahua, Chihuahua, Mexico

2 Tissue Culture Laboratory, Division of Engineering and Sciences, Monterrey Institute of Technology and Higher Education, Chihuahua, Mexico

*Address all correspondence to: eaespinoza@uach.mx

IntechOpen

© 2023 The Author(s). Licensee IntechOpen. This chapter is distributed under the terms of the Creative Commons Attribution License (http://creativecommons.org/licenses/by/3.0), which permits unrestricted use, distribution, and reproduction in any medium, provided the original work is properly cited. (cc) BY

References

[1] FAO. Plant health 2020 [Internet]. 2020. Available from: https://www.fao.org/plant-health-2020/en/. [Accesed: March 12, 2023]

[2] Liu Y-x, Li F, Gao L, Tu Z-l, Zhou F, Lin Y-j. Advancing approach and toolbox in optimization of chloroplast genetic transformation technology. Journal of Integrative Agriculture. 2023; In Press. DOI: 10.1016/j.jia.2023.02.031

[3] UN. Department of Economic and Social Affairs, Population Division, World population prospects 2022 [Internet]. 2022. Available from: https://population.un.org/wpp/. [Accesed: April 11, 2023]

[4] Xu J, Towler M, Weathers PJ. Platforms for plant-based protein production. In: Pavlov A, Bley T, editors. Bioprocessing of Plant *In Vitro* Systems. Cham: Springer International Publishing; 2018. pp. 509-548. DOI: 10.1007/978-3-319-54600-1_14

[5] Rozov SM, Sidorchuk YV, Deineko EV. Transplastomic plants: Problems of production and their solution. Russian Journal of Plant Physiology. 2022;**69**(2):20. DOI: 10.1134/S1021443722020157

[6] Kwon KC, Nityanandam R, New JS, Daniell H. Oral delivery of bioencapsulated exendin-4 expressed in chloroplasts lowers blood glucose level in mice and stimulates insulin secretion in beta-TC 6 cells. Plant Biotechnology Journal. 2012;**11**(1):77-86. DOI: 10.1111/pbi.12008

[7] Shenoy V, Kwon K-C, Rathinasabapathy A, Lin S, Jin G, Song C, et al. Oral delivery of angiotensin-converting enzyme 2 and angiotensin-(1-7) bioencapsulated in plant cells attenuates pulmonary hypertension. Hypertension. 2014;**64**(6):1248-1259. DOI: 10.1161/HYPERTENSIONAHA.114.03871

[8] Labuschagne M. A review of cereal grain proteomics and its potential for sorghum improvement. Journal of Cereal Science. 2018;**84**:151-158. DOI: 10.1016/j.jcs.2018.10.010

[9] Farquhar J, Zerkle AL, Bekker A. Geological constraints on the origin of oxygenic photosynthesis. Photosynthesis Research. 2011;**107**(1):11-36. DOI: 10.1007/s11120-010-9594-0

[10] Taylor NL, Stroher E, Millar AH. Arabidopsis organelle isolation and characterization. Methods in Molecular Biology. 2014;**1062**:551-572. DOI: 10.1007/978-1-62703-580-4_29

[11] Daniell H, Lin CS, Yu M, Chang WJ. Chloroplast genomes: Diversity, evolution, and applications in genetic engineering. Genome Biology. 2016;**17**(1):134. DOI: 10.1186/s13059-016-1004-2

[12] Singhal R, Pal R, Dutta S. Chloroplast engineering: Fundamental insights and its application in amelioration of environmental stress. Applied Biochemistry and Biotechnology. 2023;**195**(4):2463-2482. DOI: 10.1007/s12010-022-03930-8

[13] Zhang Y, Zhang A, Li X, Lu C. The role of chloroplast gene expression in plant responses to environmental stress. International Journal of Molecular Sciences. 2020;**21**(17):6082. DOI: 10.3390/ijms21176082

[14] Mamaeva A, Taliansky M, Filippova A, Love AJ, Golub N, Fesenko I. The role of chloroplast protein remodeling

in stress responses and shaping of the plant peptidome. New Phytologist. 2020;**227**(5):1326-1334. DOI: 10.1111/nph.16620

[15] Grabsztunowicz M, Koskela MM, Mulo P. Post-translational modifications in regulation of chloroplast function: Recent advances. Frontiers in Plant Science. 2017;**8**:240. DOI: 10.3389/fpls.2017.00240

[16] Kusnetsov VV. Chloroplasts: Structure and expression of the plastid genome. Russian Journal of Plant Physiology. 2018;**65**(4):465-476. DOI: 10.1134/S1021443718030044

[17] Leister D. Chloroplast research in the genomic age. TRENDS in Genetics. 2003;**19**(1):47-56. DOI: 10.1016/S0168-9525(02)00003-3

[18] Green BR. Chloroplast genomes of photosynthetic eukaryotes. The Plant Journal. 2011;**66**(1):34-44. DOI: 10.1111/j.1365-313X.2011.04541.x

[19] Gruenstaeudl M, Nauheimer L, Borsch T. Plastid genome structure and phylogenomics of Nymphaeales: Conserved gene order and new insights into relationships. Plant Systematics and Evolution. 2017;**303**(9):1251-1270. DOI: 10.1007/s00606-017-1436-5

[20] Asaf S, Khan AL, Khan AR, Waqas M, Kang S-M, Khan MA, et al. Complete chloroplast genome of *Nicotiana otophora* and its comparison with related species. Frontiers in Plant Science. 2016;7:843. DOI: 10.3389/fpls.2016.00843

[21] Heinhorst S, Cannon GC. DNA replication in chloroplasts. Journal of Cell Science. 1993;**104**(1):1-9. DOI: 10.1242/jcs.104.1.1

[22] Scott NS, Tymms MJ, Possingham JV. Plastid-DNA levels in the different tissues of potato. Planta. 1984;**161**(1):12-19. DOI: 10.1007/BF00951454

[23] Boffey SA, Leech RM. Chloroplast DNA levels and the control of chloroplast division in light-grown wheat leaves. Plant Physiology. 1982;**69**(6):1387-1391. DOI: 10.1104/pp.69.6.1387

[24] Choi IS, Jansen R, Ruhlman T. Lost and found: Return of the inverted repeat in the legume clade defined by its absence. Genome Biology and Evolution. 2019;**11**(4):1321-1333. DOI: 10.1093/gbe/evz076

[25] Palmer JD, Thompson WF. Chloroplast DNA rearrangements are more frequent when a large inverted repeat sequence is lost. Cell. 1982;**29**(2):537-550. DOI: 10.1016/0092-8674(82)90170-2

[26] Wolfe KH, Li WH, Sharp PM. Rates of nucleotide substitution vary greatly among plant mitochondrial, chloroplast, and nuclear DNAs. Proceedings of the National Academy of Sciences. 1987;**84**(24):9054-9058. DOI: 10.1073/pnas.84.24.9054

[27] Rawson JRY, Kushner SR, Vapnek D, Kirby Alton N, Boerma CL. Chloroplast ribosomal RNA genes in *Euglena gracilis* exist as three clustered tandem repeats. Gene. 1978;**3**(3):191-209. DOI: 10.1016/0378-1119(78)90032-X

[28] Kim JS, Kim HT, Kim J-H. The largest plastid genome of monocots: A novel genome type containing at residue repeats in the slipper orchid *Cypripedium japonicum*. Plant Molecular Biology Reporter. 2015;**33**(5):1210-1220. DOI: 10.1007/s11105-014-0833-y

[29] Kim JI, Shin H, Škaloud P, Jung J, Yoon HS, Archibald JM, et al. Comparative plastid genomics of Synurophyceae: Inverted repeat dynamics

and gene content variation. BMC Evolutionary Biology. 2019;**19**(1):20. DOI: 10.1186/s12862-018-1316-9

[30] Melonek J, Matros A, Trösch M, Mock HP, Krupinska K. The core of chloroplast nucleoids contains architectural SWIB domain proteins. The Plant Cell. 2012;**24**(7):3060-3073. DOI: 10.1105/tpc.112.099721

[31] Sato N, Nakayama M, Hase T. The 70-kDa major DNA-compacting protein of the chloroplast nucleoid is sulfite reductase. FEBS Letters. 2001;**487**(3):347-350. DOI: 10.1016/S0014-5793(00)02342-5

[32] Krupinska K, Desel C, Frank S, Hensel G. WHIRLIES are multifunctional DNA-binding proteins with impact on plant development and stress resistance. Frontiers in Plant Science. 2022;**13**:880423. DOI: 10.3389/fpls.2022.880423

[33] Murakami S, Kondo Y, Nakano T, Sato F. Protease activity of CND41, a chloroplast nucleoid DNA-binding protein, isolated from cultured tobacco cells. FEBS Letters. 2000;**468**(1):15-18. DOI: 10.1016/S0014-5793(00)01186-8

[34] Bohne AV. The nucleoid as a site of rRNA processing and ribosome assembly. Frontiers in Plant Science. 2014;**5**:257. DOI: 10.3389/fpls.2014.00257

[35] Melonek J, Oetke S, Krupinska K. Multifunctionality of plastid nucleoids as revealed by proteome analyses. Biochimica et Biophysica Acta (BBA) - Proteins and Proteomics. 2016;**1864**(8):1016-1038. DOI: 10.1016/j.bbapap.2016.03.009

[36] Espinoza-Sánchez EA, Torres-Castillo JA, Rascón-Cruz Q, Zavala-García F, Sinagawa-García SR. Production and characterization of fungal β-glucosidase and bacterial cellulases by tobacco chloroplast transformation. Plant Biotechnology Reports. 2016;**10**(2):61-73. DOI: 10.1007/s11816-016-0386-7

[37] Rascón-Cruz Q, González- Barriga CD, Iglesias-Figueroa BF, Trejo-Muñoz JC, Siqueiros-Cendón T, Sinagawa-García SR, et al. Plastid transformation: Advances and challenges for its implementation in agricultural crops. Electronic Journal of Biotechnology. 2021;**51**:95-109. DOI: 10.1016/j.ejbt.2021.03.005

[38] Bu Y, Sun B, Zhou A, Zhang X, Takano T, Liu S. Overexpression of *AtOxR* gene improves abiotic stresses tolerance and vitamin C content in *Arabidopsis thaliana*. Journal BMC Biotechnology. 2016;**16**(1):69. DOI: 10.1186/s12896-016-0299-0

[39] Brookes G, Barfoot P. Farm income and production impacts of using GM crop technology 1996-2016. GM Crops & Food. 2018;**9**(2):59-89. DOI: 10.1080/21645698.2018.1464866

[40] Tschofen M, Knopp D, Hood E, Stöger E. Plant molecular farming: Much more than medicines. Journal of Annual Review of Analytical Chemistry. 2016;**9**:271-294. DOI: 10.1146/annurev-anchem-071015-041706

[41] ISAAA. Global Status of Commercialized Biotech/GM Crops in 2019. In: BRIEF 55. International Service for the Acquisition of Agri-Biotech Applications. 2019

[42] USDA. Economic Research Service U.S. Department of Agriculture, Recent trends in GE adoption [Internet]. 2020. Available from: https://www.usda.gov/. [Accesed: April 13, 2023]

[43] ISAAA. GM Approval Database/event: IND-ØØ412-7 [Internet]. 2023. Available from: https://www.isaaa.org/

gmapprovaldatabase/event/default.asp?EventID=574. [Accesed: April 13, 2023]

[44] Vaeck M, Reynaerts A, Höfte H, Jansens S, De Beuckeleer M, Dean C, et al. Transgenic plants protected from insect attack. Nature. 1987;**328**(6125):33-37. DOI: 10.1038/328033a0

[45] McBride KE, Svab Z, Schaaf DJ, Hogan PS, Stalker DM, Maliga P. Amplification of a chimeric *bacillus* gene in chloroplasts leads to an extraordinary level of an insecticidal protein in tobacco. Bio/Technology. 1995;**13**(4):362-365. DOI: 10.1038/nbt0495-362

[46] Daniell H, Datta R, Varma S, Gray S, Lee S-B. Containment of herbicide resistance through genetic engineering of the chloroplast genome. Nature Biotechnology. 1998;**16**(4):345. DOI: 10.1038/nbt0498-345

[47] Songstad D, Petolino J, Voytas DF, Reichert N. Genome editing of plants. Journal of Critical Reviews in Plant Sciences. 2017;**36**(1):1-23. DOI: 10.1080/07352689.2017.1281663

[48] Rajeevkumar S, Anunanthini P, Ramalingam S. Epigenetic silencing in transgenic plants. Journal of Frontiers in Plant Science. 2015;**6**:693. DOI: 10.3389/fpls.2015.00693

[49] Dively GP, Huang F, Oyediran I, Burd T, Morsello S. Evaluation of gene flow in structured and seed blend refuge systems of non-Bt and Bt corn. Journal of Pest Science. 2020;**93**(1):439-447. DOI: 10.1007/s10681-016-1637-y

[50] Svab Z, Hajdukiewicz P, Maliga P. Stable transformation of plastids in higher plants. Proceedings of the National Academy of Sciences. 1990;**87**(21):8526-8530. DOI: 10.1073/pnas.87.21.8526

[51] Svab Z, Maliga P. High-frequency plastid transformation in tobacco by selection for a chimeric *aadA* gene. Proceedings of the National Academy of Sciences. 1993;**90**(3):913-917. DOI: 10.1073/pnas.90.3.913

[52] Davarpanah SJ, Jung SH, Kim YJ, Park Y-I, Min SR, Liu JR, et al. Stable plastid transformation in *Nicotiana benthamiana*. Journal of Plant Biology. 2009;**52**(3):244-250. DOI: 10.1007/s12374-009-9023-0

[53] Maliga P, Svab Z. Engineering the plastid genome of *Nicotiana sylvestris*, a diploid model species for plastid genetics. In: Birchler J, editor. Plant Chromosome Engineering: Methods and Protocols. Totowa, NJ: Humana Press; 2010. DOI: 10.1007/978-1-61737-957-4_2

[54] Zubko MK, Zubko EI, Kv Z, Meyer P, Day A. Stable transformation of petunia plastids. Transgenic Research. 2004;**13**(6):523-530. DOI: 10.1007/s11248-004-2374-x

[55] Singh AK, Verma SS, Bansal KC. Plastid transformation in eggplant (*Solanum melongena* L.). Transgenic Research. 2010;**19**(1):113-119. DOI: 10.1007/s11248-009-9290-z

[56] Ruf S, Hermann M, Berger IJ, Carrer H, Bock R. Stable genetic transformation of tomato plastids and expression of a foreign protein in fruit. Nature Biotechnology. 2001;**19**(9):870. DOI: 10.1038/nbt0901-870

[57] Kota S, Lakkam R, Kasula K, Narra M, Qiang H, Rao Allini V, et al. Construction of a species-specific vector for improved plastid transformation efficiency in *Capsicum annuum* L. 3 Biotech. 2019;**9**(6):226. DOI: 10.1007/s13205-019-1747-z

[58] Sidorov VA, Kasten D, Pang SZ, Hajdukiewicz PT, Staub JM, Nehra NS.

Stable chloroplast transformation in potato: Use of green fluorescent protein as a plastid marker. The Plant Journal. 1999;**19**(2):209-216. DOI: 10.1046/j.1365-313X.1999.00508.x

[59] Valkov VT, Gargano D, Manna C, Formisano G, Dix PJ, Gray JC, et al. High efficiency plastid transformation in potato and regulation of transgene expression in leaves and tubers by alternative 5′ and 3′ regulatory sequences. Transgenic Research. 2011;**20**(1):137-151. DOI: 10.1007/s11248-010-9402-9

[60] Cheng L, Li H-P, Qu B, Huang T, Tu J-X, Fu T-D, et al. Chloroplast transformation of rapeseed (*Brassica napus*) by particle bombardment of cotyledons. Journal of Plant Cell Reports. 2010;**29**(4):371-381. DOI: 10.1007/s00299-010-0828-6

[61] Liu CW, Lin CC, Chen JJ, Tseng MJ. Stable chloroplast transformation in cabbage (*Brassica oleracea* L. var. capitata L.) by particle bombardment. Plant Cell Reports. 2007;**26**(10):1733-1744. DOI: 10.1007/s00299-007-0374-z

[62] Nugent GD, Coyne S, Nguyen TT, Kavanagh TA, Dix PJ. Nuclear and plastid transformation of Brassica oleracea var. botrytis (cauliflower) using PEG-mediated uptake of DNA into protoplasts. Plant Science. 2006;**170**(1):135-142

[63] Skarjinskaia M, Svab Z, Maliga P. Plastid transformation in *Lesquerella fendleri,* an oilseed *Brassicacea*. Transgenic Research. 2003;**12**(1):115-122. DOI: 10.1023/A:1022110402302

[64] Hou B-K, Zhou Y-H, Wan L-H, Zhang Z-L, Shen G-F, Chen Z-H, et al. Chloroplast transformation in oilseed rape. Transgenic Research. 2003;**12**(1):111-114. DOI: 10.1023/A:1022180315462

[65] Sikdar S, Serino G, Chaudhuri S, Maliga P. Plastid transformation in *Arabidopsis thaliana*. Plant Cell Reports. 1998;**18**(1-2):20-24. DOI: 10.1007/s002990050525

[66] Wei Z, Liu Y, Lin C, Wang Y, Qa C, Dong Y, et al. Transformation of alfalfa chloroplasts and expression of green fluorescent protein in a forage crop. Biotechnology Letters. 2011;**33**(12):2487-2494. DOI: 10.1007/s10529-011-0709-2

[67] Dufourmantel N, Pelissier B, Garcon F, Peltier G, Ferullo JM, Tissot G. Generation of fertile transplastomic soybean. Plant Molecular Biology. 2004;**55**(4):479-489. DOI: 10.1007/s11103-004-0192-4

[68] Kumar S, Dhingra A, Daniell H. Plastid-expressed betaine aldehyde dehydrogenase gene in carrot cultured cells, roots, and leaves confers enhanced salt tolerance. Plant Physiology. 2004;**136**(1):2843-2854. DOI: 10.1104/pp.104.045187

[69] Kumar S, Dhingra A, Daniell H. Stable transformation of the cotton plastid genome and maternal inheritance of transgenes. Plant Molecular Biology. 2004;**56**(2):203-216. DOI: 10.1007/s11103-004-2907-y

[70] Kaushal C, Abdin MZ, Kumar S. Chloroplast genome transformation of medicinal plant *Artemisia annua*. Plant Biotechnology Journal. 2020;**18**(11):2155-2157. DOI: 10.1111/pbi.13379

[71] Queiroz LN, Maldaner FR, Mendes ÉA, Sousa AR, D'Allastta RC, Mendonça G, et al. Evaluation of lettuce chloroplast and soybean cotyledon as platforms for production of functional bone morphogenetic protein 2. Journal of Transgenic Research. 2019;**28**(2):213-224. DOI: 10.1007/s11248-019-00116-7

[72] Lelivelt CL, McCabe MS, Newell CA, Bastiaan deSnoo C, Van Dun KM, Birch-Machin I, et al. Stable plastid transformation in lettuce (*Lactuca sativa L.*). Plant Molecular Biology. 2005;**58**(6):763-774. DOI: 10.1007/s11103-005-7704-8

[73] Kanamoto H, Yamashita A, Asao H, Okumura S, Takase H, Hattori M, et al. Efficient and stable transformation of *Lactuca sativa* L. cv. Cisco (lettuce) plastids. Transgenic Research. 2006;**15**(2):205-217. DOI: 10.1007/s11248-005-3997-2

[74] Muralikrishna N, Srinivas K, Kumar KB, Sadanandam A. Stable plastid transformation in *Scoparia dulcis* L. Physiology and Molecular Biology of Plants. 2016;**22**(4):575-581. DOI: 10.1007/s12298-016-0386-7

[75] Narra M, Kota S, Velivela Y, Ellendula R, Allini VR, Abbagani S. Construction of chloroplast transformation vector and its functional evaluation in *Momordica charantia* L. 3 Biotech. 2018;**8**(3):140. DOI: 10.1007/s13205-018-1160-z

[76] De Marchis F, Wang Y, Stevanato P, Arcioni S, Bellucci M. Genetic transformation of the sugar beet plastome. Transgenic Research. 2008;**18**(1):17. DOI: 10.1007/s11248-008-9193-4

[77] Wu Y, Xu L, Chang L, Ma M, You L, Jiang C, et al. *Bacillus thuringiensis cry1C* expression from the plastid genome of poplar leads to high mortality of leaf-eating caterpillars. Tree Physiology. 2019;**39**(9):1525-1532. DOI: 10.1093/treephys/tpz073

[78] Okumura S, Sawada M, Park YW, Hayashi T, Shimamura M, Takase H, et al. Transformation of poplar (*Populus alba*) plastids and expression of foreign proteins in tree chloroplasts. Transgenic Research. 2006;**15**(5):637-646. DOI: 10.1007/s11248-006-9009-3

[79] Mustafa G, Khan MS. Transmission of engineered plastids in sugarcane, a C4 monocotyledonous plant, reveals that sorting of preprogrammed progenitor cells produce heteroplasmy. Plants. 2020;**10**(1):26. DOI: 10.3390/plants10010026

[80] Wang Y, Wei Z, Xing S. Stable plastid transformation of rice, a monocot cereal crop. Biochemical and biophysical research communications. 2018;**503**(4):2376-2379

[81] Lee SM, Kang K, Chung H, Yoo SH, Xu XM, Lee SB, et al. Plastid transformation in the monocotyledonous cereal crop, rice (*Oryza sativa*) and transmission of transgenes to their progeny. Molecules and Cells. 2006;**21**(3):401-410

[82] Chiyoda S, Linley PJ, Yamato KT, Fukuzawa H, Yokota A, Kohchi T. Simple and efficient plastid transformation system for the liverwort *Marchantia polymorpha* L. suspension-culture cells. Transgenic Research. 2007;**16**(1):41-49. DOI: 10.1007/s11248-006-9027-1

[83] Sugiura C, Sugita M. Plastid transformation reveals that moss tRNAArg-CCG is not essential for plastid function. The Plant Journal. 2004;**40**(2):314-321. DOI: 10.1111/j.1365-313X.2004.02202.x

[84] Bo Y, Wang K, Wu Y, Cao H, Cui Y, Wang L. Establishment of a chloroplast transformation system in *Tisochrysis lutea*. Journal of Applied Phycology. 2020;**32**(5):2959-2965. DOI: 10.1007/s10811-020-02159-4

[85] Gan Q, Jiang J, Han X, Wang S, Lu Y. Engineering the chloroplast genome of oleaginous marine microalga *Nannochloropsis oceanica*. Frontiers in plant science. 2018;**9**:439. DOI: 10.3389/fpls.2018.00439

[86] Xie W-H, Zhu C-C, Zhang N-S, Li D-W, Yang W-D, Liu J-S, et al.

Construction of novel chloroplast expression vector and development of an efficient transformation system for the diatom *Phaeodactylum tricornutum*. Marine Biotechnology. 2014;**16**(5):538-546. DOI: 10.1007/s10126-014-9570-3

[87] Zienkiewicz M, Krupnik T, Drożak A, Golke A, Romanowska E. Transformation of the *Cyanidioschyzon merolae* chloroplast genome: Prospects for understanding chloroplast function in extreme environments. Plant Molecular Biology. 2017;**93**(1):171-183. DOI: 10.1007/s11103-016-0554-8

[88] Kong F, Zhao H, Liu W, Li N, Mao Y. Construction of plastid expression vector and development of genetic transformation system for the seaweed *Pyropia yezoensis*. Marine Biotechnology. 2017;**19**(2):147-156. DOI: 10.1007/s10126-017-9736-x

[89] Lapidot M, Raveh D, Sivan A, Arad SM, Shapira M. Stable chloroplast transformation of the unicellular red alga *Porphyridium* species. Plant Physiology. 2002;**129**(1):7-12. DOI: 10.1104/pp.011023

[90] Georgianna DR, Hannon MJ, Marcuschi M, Wu S, Botsch K, Lewis AJ, et al. Production of recombinant enzymes in the marine alga *Dunaliella tertiolecta*. Algal Research. 2013;**2**(1):2-9. DOI: 10.1016/j.algal.2012.10.004

[91] Gutierrez CL, Gimpel J, Escobar C, Marshall SH, Henríquez V. Chloroplast genetic tool for the green microalgae *Haematococcus pluvialis* (Chlorophyceae, Volvocales). Journal of Phycology. 2012;**48**(4):976-983. DOI: 10.1111/j.1529-8817.2012.01178.x

[92] Doetsch NA, Favreau MR, Kuscuoglu N, Thompson MD, Hallick RB. Chloroplast transformation in *Euglena gracilis*: Splicing of a group III twintron transcribed from a transgenic psbK operon. Current Genetics. 2001;**39**(1):49-60. DOI: 10.1007/s002940000174

[93] Goldschmidt-Clermont M. Transgenic expression of aminoglycoside adenine transferase in the chloroplast: A selectable marker for site-directed transformation of *Chlamydomonas*. Journal of Nucleic Acids Research. 1991;**19**(15):4083-4089. DOI: 10.1093/nar/19.15.4083

[94] Morgenfeld MM, Vater CF, Alfano EF, Boccardo NA, Bravo-Almonacid FF. Translocation from the chloroplast stroma into the thylakoid lumen allows expression of recombinant epidermal growth factor in transplastomic tobacco plants. Transgenic Research. 2020;**29**:295-305. DOI: 10.1007/s11248-020-00199-7

[95] Albarracín RM, Becher ML, Farran I, Sander VA, Corigliano MG, Yácono ML, et al. The fusion of *toxoplasma gondii* SAG1 vaccine candidate to *Leishmania infantum* heat shock protein 83-kDa improves expression levels in tobacco chloroplasts. Biotechnology Journal. 2015;**10**(5):748-759. DOI: 10.1002/biot.201400742

[96] Su J, Zhu L, Sherman A, Wang X, Lin S, Kamesh A, et al. Low cost industrial production of coagulation factor IX bioencapsulated in lettuce cells for oral tolerance induction in hemophilia B. Biomaterials. 2015;**70**:84-93. DOI: 10.1016/j.biomaterials.2015.08.004

[97] Molina A, Hervas-Stubbs S, Daniell H, Mingo-Castel AM, Veramendi J. High-yield expression of a viral peptide animal vaccine in transgenic tobacco chloroplasts. Plant Biotechnology Journal. 2004;**2**(2):141-153. DOI: 10.1046/j.1467-7652.2004.00057.x

[98] You L, Song Q, Wu Y, Li S, Jiang C, Chang L, et al. Accumulation of glycine betaine in transplastomic potato plants expressing choline oxidase confers

improved drought tolerance. Planta. 2019;**249**(6):1963-1975. DOI: 10.1007/s00425-019-03132-3

[99] Roudsari FM, Salmanian AH, Mousavi A, Hashemi Sohi H, Jafari M. Regeneration of glyphosate-tolerant *Nicotiana tabacum* after plastid transformation with a mutated variant of bacterial *aroA* gene. Iranian Journal of Biotechnology. 2009;**7**(4):247-253

[100] Paoletti MG, Pimentel D. Environmental risks of pesticides versus genetic engineering for agricultural Pest control. Journal of Agricultural and Environmental Ethics. 2000;**12**(3):279-303. DOI: 10.1023/A:1009571131089

[101] FAO. Pesticides Indicators [Internet]. 2019. Available from: https://www.fao.org/faostat/en/#data/EP. [Accesed: March 11, 2023]

[102] APRD. Arthropod Pesticide Resistance Database (APRD) [Internet]. 2022. Available from: https://www.pesticideresistance.org/. [Accesed: March 12, 2023]

[103] Boedeker W, Watts M, Clausing P, Marquez E. The global distribution of acute unintentional pesticide poisoning: Estimations based on a systematic review. BMC Public Health. 2020;**20**(1):1875. DOI: 10.1186/s12889-020-09939-0

[104] Fauzi NIM, Fen YW, Omar NAS, Hashim HS. Recent advances on detection of insecticides using optical sensors. Sensors. 2021;**21**(11):3856. DOI: 10.3390/s21113856

[105] Olisah C, Rubidge G, Human LRD, Adams JB. A translocation analysis of organophosphate pesticides between surface water, sediments and tissues of common reed *Phragmites australis*. Chemosphere. 2021;**284**:131380. DOI: 10.1016/j.chemosphere.2021.131380

[106] Sapbamrer R, Hongsibsong S. Organophosphorus pesticide residues in vegetables from farms, markets, and a supermarket around Kwan Phayao Lake of northern Thailand. Archives of Environmental Contamination and Toxicology. 2014;**67**(1):60-67. DOI: 10.1007/s00244-014-0014-x

[107] Singsit C, Adang MJ, Lynch RE, Anderson WF, Wang A, Cardineau G, et al. Expression of a *bacillus thuringiensis cryIA (c)* gene in transgenic peanut plants and its efficacy against lesser cornstalk borer. Transgenic Research. 1997;**6**(2):169-176. DOI: 10.1023/a:1018481805928

[108] Shu Q, Ye G, Cui H, Cheng X, Xiang Y, Wu D, et al. Transgenic rice plants with a synthetic *cry1Ab* gene from *bacillus thuringiensis* were highly resistant to eight lepidopteran rice pest species. Molecular Breeding. 2000;**6**(4):433-439. DOI: 10.1023/A:1009658024114

[109] Cao J, Zhao JZ, Tang J, Shelton A, Earle E. Broccoli plants with pyramided *cry1Ac* and *cry1C* Bt genes control diamondback moths resistant to *Cry1A* and *Cry1C* proteins. Theoretical and Applied Genetics. 2002;**105**(2):258-264. DOI: 10.1007/s00122-002-0942-0

[110] Cao J, Shelton AM, Earle ED. Development of transgenic collards (*Brassica oleracea* L., var. acephala) expressing a *cry1Ac or cry1C* Bt gene for control of the diamondback moth. Crop Protection. 2005;**24**(9):804-813. DOI: 10.1016/j.cropro.2004.12.014

[111] Cao J, Shelton AM, Earle ED. Sequential transformation to pyramid two Bt genes in vegetable Indian mustard (*Brassica juncea* L.) and its potential for control of diamondback moth larvae. Plant Cell Reports. 2007;**27**(3):479. DOI: 10.1007/s00299-007-0473-x

[112] Ye W, Zhu L, Liu Y, Crickmore N, Peng D, Ruan L, et al. Mining new crystal protein genes from bacillus thuringiensis on the basis of mixed plasmid-enriched genome sequencing and a computational pipeline. Applied Environmental Microbiology. 2012;**78**(14):4795-4801. DOI: 10.1128/AEM.00340-12

[113] Guo Y, Weng M, Sun Y, Carballar-Lejarazú R, Wu S, Lian C. *Bacillus thuringiensis* toxins with nematocidal activity against the pinewood nematode *Bursaphelenchus xylophilus*. Journal of Invertebrate Pathology. 2022;**189**:107726. DOI: 10.1016/j.jip.2022.107726

[114] Welch KL, Unnithan GC, Degain BA, Wei J, Zhang J, Li X, et al. Cross-resistance to toxins used in pyramided Bt crops and resistance to Bt sprays in *Helicoverpa zea*. Journal of Invertebrate Pathology. 2015;**132**:149-156. DOI: 10.1016/j.jip.2015.10.003

[115] Tabashnik BE, Carrière Y. Surge in insect resistance to transgenic crops and prospects for sustainability. Nature Biotechnology. 2017;**35**(10):926-935. DOI: 10.1038/nbt.3974

[116] Zukoff SN, Ostlie KR, Potter B, Meihls LN, Zukoff AL, French L, et al. Multiple assays indicate varying levels of cross resistance in Cry3Bb1-selected field populations of the Western corn rootworm to mCry3A, eCry3.1Ab, and Cry34/35Ab1. Journal of Economic Entomology. 2016;**109**(3):1387-1398. DOI: 10.1093/jee/tow073

[117] Bernardi D, Salmeron E, Horikoshi RJ, Bernardi O, Dourado PM, Carvalho RA, et al. Cross-resistance between Cry1 proteins in fall armyworm (*Spodoptera frugiperda*) may affect the durability of current pyramided bt maize hybrids in Brazil. PLoS One. 2015;**10**(10):e0140130-e0140130. DOI: 10.1371/journal.pone.0140130

[118] Huang F, Buschman LL, Higgins RA, McGaughey WH. Inheritance of resistance to *bacillus thuringiensis* toxin (Dipel ES) in the european corn borer. Science. 1999;**284**(5416):965-967. DOI: 10.1126/science.284.5416.965

[119] Liu C-W, Lin C-C, Yiu J-C, Chen JJW, Tseng M-J. Expression of a *bacillus thuringiensis* toxin (*cry1Ab*) gene in cabbage (*Brassica oleracea* L. var. capitata L.) chloroplasts confers high insecticidal efficacy against *Plutella xylostella*. Theoretical and Applied Genetics. 2008;**117**(1):75-88. DOI: 10.1007/s00122-008-0754-y

[120] De Cosa B, Moar W, Lee SB, Miller M, Daniell H. Overexpression of the Bt cry2Aa2 operon in chloroplasts leads to formation of insecticidal crystals. Nature Biotechnology. 2001;**19**(1):71-74. DOI: 10.1038/83559

[121] Chakrabarti SK, Lutz KA, Lertwiriyawong B, Svab Z, Maliga P. Expression of the *cry9Aa2* B.t. gene in tobacco chloroplasts confers resistance to potato tuber moth. Transgenic Research. 2006;**15**(4):481-488. DOI: 10.1007/s11248-006-0018-z

[122] Dufourmantel N, Tissot G, Goutorbe F, Garçon F, Muhr C, Jansens S, et al. Generation and analysis of soybean plastid transformants expressing *bacillus thuringiensis* Cry1Ab protoxin. Plant Molecular Biology. 2005;**58**(5):659-668. DOI: 10.1007/s11103-005-7405-3

[123] Reddy VS, Leelavathi S, Selvapandiyan A, Raman R, Giovanni F, Shukla V, et al. Analysis of chloroplast transformed tobacco plants with cry1Ia5 under rice psbA transcriptional elements reveal high level expression of Bt toxin without imposing yield penalty and stable inheritance of transplastome. Molecular Breeding. 2002;**9**(4):259-269. DOI: 10.1023/A:1020357729437

[124] Muzaffar A, Kiani S, Khan MAU, Rao AQ, Ali A, Awan MF, et al. Chloroplast localization of Cry1Ac and Cry2A protein-an alternative way of insect control in cotton. Biological Research. 2015;**48**(1):1-11. DOI: 10.1186/s40659-015-0005-z

[125] Wu J, Luo X, Zhang X, Shi Y, Tian Y. Development of insect-resistant transgenic cotton with chimeric TVip3A* accumulating in chloroplasts. Transgenic Research. 2011;**20**(5):963-973. DOI: 10.1007/s11248-011-9483-0

[126] Kiani S, Mohamed BB, Shehzad K, Jamal A, Shahid MN, Shahid AA, et al. Chloroplast-targeted expression of recombinant crystal-protein gene in cotton: An unconventional combat with resistant pests. Journal of Biotechnology. 2013;**166**(3):88-96. DOI: 10.1016/j.jbiotec.2013.04.011

[127] Li X, Li S, Lang Z, Zhang J, Zhu L, Huang D. Chloroplast-targeted expression of the codon-optimized truncated *cry1Ah* gene in transgenic tobacco confers a high level of protection against insects. Plant Cell Reports. 2013;**32**(8):1299-1308. DOI: 10.1007/s00299-013-1444-z

[128] Wang Y-P, Wei Z-Y, Zhang Y-Y, Lin C-J, Zhong X-F, Wang Y-L, et al. Chloroplast-expressed MSI-99 in tobacco improves disease resistance and displays inhibitory effect against rice blast fungus. International Journal of Molecular Sciences. 2015;**16**(3):4628-4641. DOI: 10.3390/ijms16034628

[129] Jin S, Zhang X, Daniell H. *Pinellia ternata* agglutinin expression in chloroplasts confers broad spectrum resistance against aphid, whitefly, lepidopteran insects, bacterial and viral pathogens. Plant Biotechnology Journal. 2011;**10**(3):313-327. DOI: 10.1111/j.1467-7652.2011.00663.x

[130] Ruhlman TA, Rajasekaran K, Cary JW. Expression of chloroperoxidase from *pseudomonas pyrrocinia* in tobacco plastids for fungal resistance. Plant Science. 2014;**228**:98-106. DOI: 10.1016/j.plantsci.2014.02.008

[131] Sauka DH, Peralta C, Pérez MP, Onco MI, Fiodor A, Caballero J, et al. *Bacillus toyonensis* biovar Thuringiensis: A novel entomopathogen with insecticidal activity against lepidopteran and coleopteran pests. Biological Control. 2022;**167**:104838. DOI: 10.1016/j.biocontrol.2022.104838

[132] Aasen SS, Hågvar EB. Effect of potato plants expressing snowdrop lectin (GNA) on the performance and colonization behaviour of the peach-potato aphid *Myzus persicae*. Journal of Acta Agriculturae Scandinavica, Section B-Soil Plant Science. 2012;**62**(4):352-361. DOI: 10.1080/09064710.2011.619996

[133] Wu K-M, Lu Y-H, Feng H-Q, Jiang Y-Y, Zhao J-Z. Suppression of cotton bollworm in multiple crops in China in areas with bt toxin-containing cotton. Science. 2008;**321**(5896):1676-1678. DOI: 10.1126/science.1160550

[134] Lu Y, Wu K, Jiang Y, Xia B, Li P, Feng H, et al. Mirid bug outbreaks in multiple crops correlated with wide-scale adoption of bt cotton in China. Science. 2010;**328**(5982):1151-1154. DOI: 10.1126/science.1187881

[135] Gq S, Sink KC, Walworth AE, Cook MA, Allison RF, Lang GA. Engineering cherry rootstocks with resistance to Prunus necrotic ring spot virus through RNA i-mediated silencing. Plant Biotechnology Journal. 2013;**11**(6):702-708. DOI: 10.1111/pbi.12060

[136] Bolognesi R, Ramaseshadri P, Anderson J, Bachman P, Clinton W,

Flannagan R, et al. Characterizing the mechanism of action of double-stranded RNA activity against western corn rootworm (*Diabrotica virgifera virgifera* LeConte). PLoS One. 2012;**7**(10):e47534. DOI: 10.1371/journal.pone.0047534

[137] Mao Y-B, Tao X-Y, Xue X-Y, Wang L-J, Chen X-Y. Cotton plants expressing CYP6AE14 double-stranded RNA show enhanced resistance to bollworms. Transgenic Research. 2011;**20**(3):665-673. DOI: 10.1007/s11248-010-9450-1

[138] Wu M, Dong Y, Zhang Q, Li S, Chang L, Loiacono FV, et al. Efficient control of western flower thrips by plastid-mediated RNA interference. Proceedings of the National Academy of Sciences. 2022;**119**(15):e2120081119. DOI: 10.1073/pnas.21200811

[139] Gordon KHJ, Waterhouse PM. RNAi for insect-proof plants. Nature Biotechnology. 2007;**25**(11):1231-1232. DOI: 10.1038/nbt1107-1231

[140] Zhang J, Khan SA, Hasse C, Ruf S, Heckel DG, Bock R. Full crop protection from an insect pest by expression of long double-stranded RNAs in plastids. Science. 2015;**347**(6225):991-994. DOI: 10.1126/science.1261680

[141] Jin S, Singh ND, Li L, Zhang X, Daniell H. Engineered chloroplast dsRNA silences cytochrome p450 monooxygenase, V-ATPase and chitin synthase genes in the insect gut and disrupts *Helicoverpa armigera* larval development and pupation. Plant Biotechnology Journal. 2015;**13**(3):435-446. DOI: 10.1111/pbi.12355

[142] Burke W, Kaplanoglu E, Kolotilin I, Menassa R, Donly C. RNA interference in the tobacco hornworm, *Manduca sexta*, using plastid-encoded long double-stranded RNA. Frontiers in Plant Science. 2019;**10**:313. DOI: 10.3389/fpls.2019.00313

[143] Bally J, McIntyre GJ, Doran RL, Lee K, Perez A, Jung H, et al. In-plant protection against *Helicoverpa armigera* by production of long hpRNA in chloroplasts. Frontiers in Plant Science. 2016;**7**:1453. DOI: 10.3389/fpls.2016.01453

[144] He W, Xu W, Xu L, Fu K, Guo W, Bock R, et al. Length-dependent accumulation of double-stranded RNAs in plastids affects RNA interference efficiency in the Colorado potato beetle. Journal of Experimental Botany. 2020;**71**(9):2670-2677. DOI: 10.1093/jxb/eraa001

[145] Turner C, Davy M, MacDiarmid R, Plummer K, Birch N, Newcomb R. RNA interference in the light brown apple moth, *Epiphyas postvittana* (Walker) induced by double-stranded RNA feeding. Insect Molecular Biology. 2006;**15**(3):383-391. DOI: 10.1111/j.1365-2583.2006.00656.x

[146] Baum JA, Bogaert T, Clinton W, Heck GR, Feldmann P, Ilagan O, et al. Control of coleopteran insect pests through RNA interference. Nature Biotechnology. 2007;**25**(11):1322-1326. DOI: 10.1038/nbt1359

[147] Ren B, Cao J, He Y, Yang S, Zhang J. Assessment on effects of transplastomic potato plants expressing Colorado potato beetle β-actin double-stranded RNAs for three non-target pests. Pesticide Biochemistry and Physiology. 2021;**178**:104909. DOI: 10.1016/j.pestbp.2021.104909

[148] Niu J, Yang W-J, Tian Y, Fan J-Y, Ye C, Shang F, et al. Topical dsRNA delivery induces gene silencing and mortality in the pea aphid. Pest Management Science. 2019;**75**(11):2873-2881. DOI: 10.1002/ps.5457

[149] Dong Y, Yang Y, Wang Z, Wu M, Fu J, Guo J, et al. Inaccessibility to double-stranded RNAs in plastids

restricts RNA interference in *Bemisia tabaci* (whitefly). Pest Management Science. 2020;**76**(9):3168-3176. DOI: 10.1002/ps.5871

[150] ISAAA. GM Approval Database [Internet]. 2023. Available from: https://www.isaaa.org/gmapprovaldatabase/. [Accesed: March 12, 2023]

[151] EPA. U.S. Environmental Protection Agency, Registers Innovative Tool to Control Corn Rootworm [Internet]. 2017. Available from: https://www.epa.gov/. [Accesed: March 17, 2023]

[152] Head GP, Carroll MW, Evans SP, Rule DM, Willse AR, Clark TL, et al. Evaluation of SmartStax and SmartStax PRO maize against western corn rootworm and northern corn rootworm: Efficacy and resistance management. Pest Management Science. 2017;**73**(9):1883-1899. DOI: 10.1002/ps.4554

[153] Mittler R. Oxidative stress, antioxidants and stress tolerance. Trends in Plant Science. 2002;7(9):405-410. DOI: 10.1016/s1360-1385(02)02312-9

[154] Zhang Y, Yu H, Yang X, Li Q, Ling J, Wang H, et al. CsWRKY46, a WRKY transcription factor from cucumber, confers cold resistance in transgenic-plant by regulating a set of cold-stress responsive genes in an ABA-dependent manner. Plant Physiology and Biochemistry. 2016;**108**:478-487. DOI: 10.1016/j.plaphy.2016.08.013

[155] Miao R, Zang W, Yuan Y, Zhang Y, Zhang A, Pang Q. The halophyte gene *ScVTC2* confers resistance to oxidative stress via AsA-mediated photosynthetic enhancement. Plant Physiology and Biochemistry. 2021;**169**:138-148. DOI: 10.1016/j.plaphy.2021.11.020

[156] Fang X, Li W, Yuan H, Chen H, Bo C, Ma Q, et al. Mutation of ZmWRKY86 confers enhanced salt stress tolerance in maize. Plant Physiology and Biochemistry. 2021;**167**:840-850. DOI: 10.1016/j.plaphy.2021.09.010

[157] Luo P, Chen Y, Rong K, Lu Y, Wang N, Xu Z, et al. *ZmSNAC13*, a maize NAC transcription factor conferring enhanced resistance to multiple abiotic stresses in transgenic *Arabidopsis*. Plant Physiology and Biochemistry. 2022;**170**:160-170. DOI: 10.1016/j.plaphy.2021.11.032

[158] Patel MK, Pandey S, Patel J, Mishra A. A type 2 metallothionein (*SbMT-2*) gene cloned from *Salicornia brachiata* confers enhanced Zn stress-tolerance in transgenic tobacco by transporting Zn2+ and maintaining photosynthesis efficacy. Environmental and Experimental Botany. 2021;**191**:104626. DOI: 10.1016/j.envexpbot.2021.104626

[159] Tiwari P, Indoliya Y, Chauhan AS, Pande V, Chakrabarty D. Over-expression of rice R1-type MYB transcription factor confers different abiotic stress tolerance in transgenic *Arabidopsis*. Ecotoxicology and Environmental Safety. 2020;**206**:111361. DOI: 10.1016/j.ecoenv.2020.111361

[160] Zhang Y, Yang L, Zhang M, Yang J, Cui J, Hu H, et al. CfAPX, a cytosolic ascorbate peroxidase gene from Cryptomeria fortunei, confers tolerance to abiotic stress in transgenic Arabidopsis. Plant Physiology and Biochemistry. 2022;**172**:167-179. DOI: 10.1016/j.plaphy.2022.01.011

[161] Torti P, Raineri J, Mencia R, Campi M, Gonzalez DH, Welchen E. The sunflower TLDc-containing protein HaOXR2 confers tolerance to oxidative stress and waterlogging when expressed in maize plants. Plant Science. 2020;**300**:110626. DOI: 10.1016/j.plantsci.2020.110626

[162] Le Martret B, Poage M, Shiel K, Nugent GD, Dix PJ. Tobacco chloroplast transformants expressing genes encoding dehydroascorbate reductase, glutathione reductase, and glutathione-S-transferase, exhibit altered anti-oxidant metabolism and improved abiotic stress tolerance. Plant Biotechnology Journal. 2011;**9**(6):661-673. DOI: 10.1111/j.1467-7652.2011.00611.x

[163] Zhang J, Tan W, Yang X-H, Zhang H-X. Plastid-expressed choline monooxygenase gene improves salt and drought tolerance through accumulation of glycine betaine in tobacco. Plant Cell Reports. 2008;**27**(6):1113. DOI: 10.1007/s00299-008-0549-2

[164] Ceccoli RD, Blanco NE, Segretin ME, Melzer M, Hanke GT, Scheibe R, et al. Flavodoxin displays dose-dependent effects on photosynthesis and stress tolerance when expressed in transgenic tobacco plants. Planta. 2012;**236**(5):1447-1458. DOI: 10.1007/s00425-012-1695-x

[165] Khan MS, Kanwal B, Nazir S. Metabolic engineering of the chloroplast genome reveals that the yeast *ArDH* gene confers enhanced tolerance to salinity and drought in plants. Frontiers in Plant Science. 2015;**6**:725. DOI: 10.3389/fpls.2015.00725

[166] Bansal K, Singh A, Wani S. Plastid transformation for abiotic stress tolerance in plants. In: Shabala S, Cuin T, editors. Plant Salt Tolerance. Totowa, NJ: Humana Press; 2012. pp. 351-358. DOI: 10.1007/978-1-61779-986-0_23

[167] Lu Y, Rijzaani H, Karcher D, Ruf S, Bock R. Efficient metabolic pathway engineering in transgenic tobacco and tomato plastids with synthetic multigene operons. Proceedings of the National Academy of Sciences. 2013;**110**(8):E623-E632. DOI: 10.1073/pnas.1216898110

[168] Poage M, Le Martret B, Jansen MAK, Nugent GD, Dix PJ. Modification of reactive oxygen species scavenging capacity of chloroplasts through plastid transformation. Plant Molecular Biology. 2011;**76**(3):371-384. DOI: 10.1007/s11103-011-9784-y

[169] Stavridou E, Michailidis M, Gedeon S, Ioakeim A, Kostas S, Chronopoulou E, et al. Tolerance of transplastomic tobacco plants overexpressing a theta class glutathione transferase to abiotic and oxidative stresses. Frontiers in Plant Science. 2019;**9**:1861. DOI: 10.3389/fpls.2018.01861

[170] Pelletier G, Budar F. The molecular biology of cytoplasmically inherited male sterility and prospects for its engineering. Current Opinion in Biotechnology. 2007;**18**(2):121-125. DOI: 10.1016/j.copbio.2006.12.002

[171] Ruiz ON, Daniell H. Engineering cytoplasmic male sterility via the chloroplast genome by expression of β-ketothiolase. Plant Physiology. 2005;**138**(3):1232-1246. DOI: 10.1104/pp.104.057729

[172] Saini JK, Saini R, Tewari L. Lignocellulosic agriculture wastes as biomass feedstocks for second-generation bioethanol production: concepts and recent developments. 3 Biotech. 2015;**5**(4):337-353. DOI: 10.1007/s13205-014-0246-5

[173] Uday USP, Choudhury P, Bandyopadhyay TK, Bhunia B. Classification, mode of action and production strategy of xylanase and its application for biofuel production from water hyacinth. International Journal of Biological Macromolecules. 2016;**82**:1041-1054. DOI: 10.1016/j.ijbiomac.2015.10.086

[174] Yan XL, Yanan W, Hongxing L, Zhigang L, Lin Z, Yinbo Q, et al. Cow

manure as a lignocellulosic substrate for fungal cellulase expression and bioethanol production. AMB Express. 2018;**8**(1):190. DOI: 10.1186/s13568-018-0720-2

[175] Bhurat KS, Banerjee T, Pandey JK, Belapurkar P. Fermentative bio-hydrogen production using lignocellulosic waste biomass: A review. Waste Disposal & Sustainable Energy. 2020;**2**:249-264. DOI: 10.1007/s42768-020-00054-9

[176] OECD/FAO. OECD-FAO Agricultural Outlook 2021-2030. Paris: OECD Publishing; 2021. p. 337. DOI: 10.1787/19428846-en

[177] Bajpai P. Pretreatment of lignocelluloses biomass for bioethanol production. In: Bajpai P, editor. Developments in Bioethanol. Singapore: Springer Singapore; 2021. pp. 111-144. DOI: 10.1007/978-981-15-8779-5_6

[178] Benedetti M, Vecchi V, Betterle N, Natali A, Bassi R, Dall'Osto L. Design of a highly thermostable hemicellulose-degrading blend from *Thermotoga neapolitana* for the treatment of lignocellulosic biomass. Journal of Biotechnology. 2019;**296**:42-52. DOI: 10.1016/j.jbiotec.2019.03.005

[179] Sanderson K. Lignocellulose: A chewy problem. Nature. 2011;**474**(7352):S12-S14. DOI: 10.1038/474S012a

[180] Mota TR, de Oliveira DM, Marchiosi R, Ferrarese-Filho O, dos Santos WD. Plant cell wall composition and enzymatic deconstruction. AIMS Bioengineering. 2018;**5**(1):63. DOI: 10.3934/bioeng.2018.1.63

[181] Zhang L, Loh KC, Zhang J. Enhanced biogas production from anaerobic digestion of solid organic wastes: Current status and prospects. Bioresource Technology Reports. 2019;**5**:280-296. DOI: 10.1016/j.biteb.2018.07.005

[182] Oliveira DM, Mota TR, Grandis A, de Morais GR, de Lucas RC, Polizeli ML, et al. Lignin plays a key role in determining biomass recalcitrance in forage grasses. Renewable Energy. 2020;**147**:2206-2217. DOI: 10.1016/j.renene.2019.10.020

[183] Park SH, Ong RG, Sticklen M. Strategies for the production of cell wall-deconstructing enzymes in lignocellulosic biomass and their utilization for biofuel production. Plant Biotechnology Journal. 2015;**14**(6):1329-1344. DOI: 10.1111/pbi.12505

[184] Hemansi GR, Kuhad RC, Saini JK. Cost effective production of complete cellulase system by newly isolated *aspergillus Niger* RCKH-3 for efficient enzymatic saccharification: Medium engineering by overall evaluation criteria approach (OEC). Biochemical Engineering Journal. 2018;**132**:182-190. DOI: 10.1016/j.bej.2018.01.019

[185] Scheller HV, Ulvskov P. Hemicelluloses. Annual Reviews. 2010;**61**(1):263-289. DOI: 10.1146/annurev-arplant-042809-112315

[186] Momeni MH, Ubhayasekera W, Sandgren M, Ståhlberg J, Hansson H. Structural insights into the inhibition of cellobiohydrolase Cel7A by xylo-oligosaccharides. The FEBS Journal. 2015;**282**(11):2167-2177. DOI: 10.1111/febs.13265

[187] Verma D, Kanagaraj A, Jin S, Singh ND, Kolattukudy PE, Daniell H. Chloroplast-derived enzyme cocktails hydrolyse lignocellulosic biomass and release fermentable sugars. Plant biotechnology journal. 2010;**8**(3):332-350. DOI: 10.1111/j.1467-7652.2009.00486.x

[188] Espinoza-Sánchez EA, Álvarez-Hernández MH, Torres-Castillo JA, Rascón-Cruz Q, Gutiérrez-Díez A, Zavala-García F, et al. Stable expression

and characterization of a fungal pectinase and bacterial peroxidase genes in tobacco chloroplast. Electronic Journal of Biotechnology. 2015;**18**(3):161-168. DOI: 10.1016/j.ejbt.2015.03.002

[189] Verma D, Jin S, Kanagaraj A, Singh ND, Daniel J, Kolattukudy PE, et al. Expression of fungal cutinase and swollenin in tobacco chloroplasts reveals novel enzyme functions and/or substrates. PLoS One. 2013;**8**(2):e57187. DOI: 10.1371/journal.pone.0057187

[190] Davarpanah S, Ahn J-W, Ko S, Jung S, Park Y-I, Liu J, et al. Stable expression of a fungal laccase protein using transplastomic tobacco. Plant Biotechnology Reports. 2012;**6**(4):305-312. DOI: 10.1007/s11816-012-0225-4

[191] Leelavathi S, Gupta N, Maiti S, Ghosh A, Siva RV. Overproduction of an alkali-and thermo-stable xylanase in tobacco chloroplasts and efficient recovery of the enzyme. Molecular Breeding. 2003;**11**:59-67

[192] Kim JY, Kavas M, Fouad WM, Nong G, Preston JF, Altpeter F. Production of hyperthermostable GH10 xylanase *Xyl10B* from *Thermotoga maritima* in transplastomic plants enables complete hydrolysis of methylglucuronoxylan to fermentable sugars for biofuel production. Plant Molecular Biology. 2011;**76**:357-369. DOI: 10.1007/s11103-010-9712-6

[193] Kolotilin I, Kaldis A, Pereira EO, Laberge S, Menassa R. Optimization of transplastomic production of hemicellulases in tobacco: Effects of expression cassette configuration and tobacco cultivar used as production platform on recombinant protein yields. Biotechnology for Biofuels. 2013;**6**(1):1-15

[194] Castiglia D, Sannino L, Marcolongo L, Ionata E, Tamburino R, De Stradis A, et al. High-level expression of thermostable cellulolytic enzymes in tobacco transplastomic plants and their use in hydrolysis of an industrially pretreated *Arundo donax* L. biomass. Biotechnology for Biofuels. 2016;**9**:1-16. DOI: 10.1186/s13068-016-0569-z

[195] Yu L-X, Gray BN, Rutzke CJ, Walker LP, Wilson DB, Hanson MR. Expression of thermostable microbial cellulases in the chloroplasts of nicotine-free tobacco. Journal of Biotechnology. 2007;**131**(3):362-369. DOI: 10.1016/j.jbiotec.2007.07.942

[196] Jin S, Kanagaraj A, Verma D, Lange T, Daniell H. Release of hormones from conjugates: Chloroplast expression of β-glucosidase results in elevated phytohormone levels associated with significant increase in biomass and protection from aphids or whiteflies conferred by sucrose esters. Plant Physiology. 2011;**155**(1):222-235. DOI: 10.1104/pp.110.160754

[197] Petersen K, Bock R. High-level expression of a suite of thermostable cell wall-degrading enzymes from the chloroplast genome. Plant Molecular Biology. 2011;**76**:311-321

[198] Gray BN, Yang H, Ahner BA, Hanson MR. An efficient downstream box fusion allows high-level accumulation of active bacterial beta-glucosidase in tobacco chloroplasts. Plant molecular biology. 2011;**76**:345-355. DOI: 10.1007/s11103-011-9743-7

[199] Nakahira Y, Ishikawa K, Tanaka K, Tozawa Y, Shiina T. Overproduction of hyperthermostable β-1, 4-endoglucanase from the archaeon *Pyrococcus horikoshii* by tobacco chloroplast engineering. Bioscience, Biotechnology, and Biochemistry. 2013;**77**(10):2140-2143. DOI: 10.1271/bbb.130413

[200] Longoni P, Leelavathi S, Doria E, Reddy VS, Cella R. Production by tobacco

transplastomic plants of recombinant fungal and bacterial cell-wall degrading enzymes to be used for cellulosic biomass saccharification. BioMed Research International. 2015;**2015**:289759. DOI: 10.1155/2015/289759

[201] Martinez D, Berka RM, Henrissat B, Saloheimo M, Arvas M, Baker SE, et al. Genome sequencing and analysis of the biomass-degrading fungus *Trichoderma reesei* (syn. *Hypocrea jecorina*). Nature Biotechnology. 2008;**26**(5):553-560. DOI: 10.1038/nbt1403

[202] Beguin P, Aubert JP. The biological degradation of cellulose. FEMS Microbiology Reviews. 1994;**13**(1):25-58. DOI: 10.1016/0168-6445(94)90099-X

[203] Wilson DB. Cellulases and biofuels. Current Opinion in Biotechnology. 2009;**20**(3):295-299. DOI: 10.1016/j.copbio.2009.05.007

[204] Liu G, Zhang J, Bao J. Cost evaluation of cellulase enzyme for industrial-scale cellulosic ethanol production based on rigorous Aspen plus modeling. Bioprocess and Biosystems Engineering. 2016;**39**(1):133-140. DOI: 10.1007/s00449-015-1497-1

[205] Gaobotse G, Venkataraman S, Mmereke KM, Moustafa K, Hefferon K, Makhzoum A. Recent Progress on vaccines produced in transgenic plants. Vaccine. 2022;**10**(11):1861. DOI: 10.3390/vaccines10111861

[206] Kwon K-C, Daniell H. Low-cost oral delivery of protein drugs bioencapsulated in plant cells. Plant Biotechnology Journal. 2015;**13**(8):1017-1022. DOI: 10.1111/pbi.12462

[207] Walsh G. Biopharmaceutical benchmarks 2014. Nature Biotechnology. 2014;**32**(10):992-1000. DOI: 10.1038/nbt.3040

[208] BIOPHARMA. Biopharmaceutical Products in the U.S. and European Markets [Internet]. 2023. Available from: http://www.biopharma.com/approvals.html. [Accesed: April 12, 2023]

[209] Kumar AU, Ling APK. Gene introduction approaches in chloroplast transformation and its applications. Journal of Genetic Engineering and Biotechnology. 2021;**19**(1):1-10. DOI: 10.1186/s43141-021-00255-7

[210] Fernández-San Millán A, Mingo-Castel A, Miller M, Daniell H. A chloroplast transgenic approach to hyper-express and purify Human serum albumin, a protein highly susceptible to proteolytic degradation. Plant Biotechnology Journal. 2003;**1**(2):71-79. DOI: 10.1046/j.1467-7652.2003.00008.x

[211] Shil PK, Kwon K-C, Zhu P, Verma A, Daniell H, Li Q. Oral delivery of ACE2/Ang-(1-7) bioencapsulated in plant cells protects against experimental uveitis and autoimmune uveoretinitis. Molecular Therapy. 2014;**22**(12):2069-2082. DOI: 10.1038/mt.2014.179

[212] Daniell H, Lee S-B, Panchal T, Wiebe PO. Expression of the native cholera toxin B subunit gene and assembly as functional oligomers in transgenic tobacco chloroplasts. Journal of Molecular Biology. 2001;**311**(5):1001-1009. DOI: 10.1006/jmbi.2001.4921

[213] Rigano MM, Manna C, Giulini A, Pedrazzini E, Capobianchi M, Castilletti C, et al. Transgenic chloroplasts are efficient sites for high-yield production of the vaccinia virus envelope protein A27L in plant cells. Plant Biotechnology Journal. 2009;**7**(6):577-591. DOI: 10.1111/j.1467-7652.2009.00425.x

[214] Lakshmi PS, Verma D, Yang X, Lloyd B, Daniell H. Low cost tuberculosis vaccine antigens in capsules: Expression

in chloroplasts, bio-encapsulation, stability and functional evaluation *in vitro*. PLoS One. 2013;**8**(1):e54708. DOI: 10.1371/journal.pone.0054708

[215] Saba K, Gottschamel J, Younus I, Syed T, Gull K, Lössl AG, et al. Chloroplast-based inducible expression of ESAT-6 antigen for development of a plant-based vaccine against tuberculosis. Journal of Biotechnology. 2019;**305**:1-10. DOI: 10.1016/j.jbiotec.2019.08.016

[216] Sherman A, Su J, Lin S, Wang X, Herzog RW, Daniell H. Suppression of inhibitor formation against FVIII in a murine model of hemophilia a by oral delivery of antigens bioencapsulated in plant cells. Blood, The Journal of the American Society of Hematology. 2014;**124**(10):1659-1668. DOI: 10.1182/blood-2013-10-528737

[217] Morgenfeld M, Lentz E, Segretin ME, Alfano EF, Bravo-Almonacid F. Translational fusion and redirection to thylakoid lumen as strategies to enhance accumulation of Human papillomavirus E7 antigen in tobacco chloroplasts. Molecular Biotechnology. 2014;**56**(11):1021-1031. DOI: 10.1007/s12033-014-9781-x

[218] Gottschamel J, Lössl A, Ruf S, Wang Y, Skaugen M, Bock R, et al. Production of dengue virus envelope protein domain III-based antigens in tobacco chloroplasts using inducible and constitutive expression systems. Plant Molecular Biology. 2016;**91**:497-512. DOI: 10.1007/s11103-016-0484-5

[219] Rosales-Mendoza S, Monreal-Escalante E, González-Ortega O, Hernández M, Fragoso G, Garate T, et al. Transplastomic plants yield a multicomponentvaccineagainstcysticercosis. Journal of Biotechnology. 2018;**266**:124-132. DOI: 10.1016/j.jbiotec.2017.12.012

[220] Hoelscher M, Tiller N, Teh AY-H, Wu G-Z, Ma JK, Bock R. High-level expression of the HIV entry inhibitor griffithsin from the plastid genome and retention of biological activity in dried tobacco leaves. Plant Molecular Biology. 2018;**97**:357-370. DOI: 10.1007/s11103-018-0744-7

[221] Tien N-Q-D, Huy N-X, Kim M-Y. Improved expression of porcine epidemic diarrhea antigen by fusion with cholera toxin B subunit and chloroplast transformation in *Nicotiana tabacum*. Plant Cell, Tissue and Organ Culture (PCTOC). 2019;**137**:213-223. DOI: 10.1007/s11240-019-01562-1

[222] Davoodi-Semiromi A, Samson N, Daniell H. The green vaccine: A global strategy to combat infectious and autoimmune diseases. Human Vaccines. 2009;**5**(7):488-493. DOI: 10.4161/hv.8247

[223] Chan HT, Daniell H. Plant-made oral vaccines against human infectious diseases-are we there yet? Plant Biotechnol Journal. 2015;**13**(8):1056-1070. DOI: 10.1111/pbi.12471

[224] Su J, Sherman A, Doerfler PA, Byrne BJ, Herzog RW, Daniell H. Oral delivery of acid alpha glucosidase epitopes expressed in plant chloroplasts suppresses antibody formation in treatment of Pompe mice. Plant Biotechnology Journal. 2015;**13**(8):1023-1032. DOI: 10.1111/pbi.12413

Chapter 5

Chloroplast Recycling and Plant Stress Tolerance

Faiz Ahmad Joyia, Ghulam Mustafa and Muhammad Sarwar Khan

Abstract

Plastids have emerged as pivotal regulators of plant's response to biotic and abiotic stresses. Chloroplasts have the ability to synthesize a variety of pigments, secondary metabolites, and phytohormones which help plant cells to withstand adverse conditions. Further, plastids communicate with the nucleus and other cellular organelles for the acquisition of essential molecules to survive under unfavorable conditions. They act as environmental sensors which not only synthesize molecules for stress tolerance but also induce nucleus-encoded genes for stress resilience. Senescence is a key developmental process in this context and plays an important role in the release of essential nutrients. Chloroplast proteolytic machinery plays a crucial role in the degradation or remodeling of plastid proteins resulting in the generation of numerous endogenous peptides which are present in the plant secretome. Plastid chaperone system is also activated for the repair/refold of damaged proteins resulting in improved tolerance to stresses. Autophagy is a conserved process that involves large-scale breakdown of chloroplast through piecemeal degradation and chlorophagy. The piecemeal degradation occurs through Rubisco-containing bodies (RCBs) and senescence-associated vacuoles (SAVs), whereas chlorophagy targets chloroplasts as a whole. Though information about chloroplast recycling is limited, the present work provides a comprehensive review on chloroplast recycling and its role in stress mitigation and adaptation in climate change scenarios.

Keywords: chloroplast proteolytic machinery, leaf senescence, chlorophagy, autophagy, stress tolerance

1. Introduction

Leaf senescence is a complex biological process characterized by the active breakdown of cellular macromolecules and the subsequent remobilization of their components. During this process, proteins like Rubisco and other chloroplast proteins undergo gradual degradation, serving as a vital source of nitrogen for recycling. This degradation is closely linked with a decline in photosynthetic activity [1]. Notably, under conditions of darkness-induced starvation, chloroplast proteins, including Rubisco, can be degraded with their carbon skeletons being utilized for respiration [2]. As the levels of these essential components diminish, chloroplasts undergo structural changes, eventually transitioning into gerontoplasts. In gerontoplasts, thylakoid membranes break down, plastoglobules accumulate, and the overall population of chloroplasts in the cell declines.

The vacuole within leaf mesophyll cells, constituting a substantial portion of the total cell volume, is rich in a diverse array of lytic hydrolases. A significant proportion of the proteolytic activity responsible for degrading Rubisco is localized within the vacuolar fraction. Notably, senescence-associated induction of various vacuolar cysteine proteases has been observed [3]. Previous studies have proposed that sequential degradation of chloroplasts within the vacuole serves as the primary pathway for chloroplast protein degradation during leaf senescence. In electron microscopy observations, chloroplasts were found to either reside within the vacuole or within tonoplast invaginations in mesophyll protoplasts from senescing wheat leaves, suggesting a mechanism resembling phagocytosis [1]. This process of delivering cytoplasmic components to the vacuole for degradation is now recognized as autophagy.

Interestingly, the decline in Rubisco protein levels occurs more rapidly than the decrease in the chloroplast population size during senescence. Furthermore, the reduction in major chloroplast proteins does not proceed uniformly; for instance, Rubisco declines more quickly than light-harvesting complex II (LHCII) [4]. Both of these proteins are primarily synthesized during leaf expansion and exhibit lower turnover rates thereafter. These observations suggest the existence of alternative pathways or mechanisms beyond whole chloroplast autophagy that contribute to Rubisco degradation. Consequently, the study of Rubisco degradation has largely focused on chloroplast proteases rather than autophagy [5]. Notably, several proteases within chloroplasts have crucial roles in chloroplast development and maintenance. Genome-wide studies have uncovered prokaryotic-origin proteases within chloroplasts, with ATP-dependent proteases such as Clp, FtsH, and Lon being prominent enzymes involved in gradual protein degradation into amino acids and oligopeptides [6]. While some of these proteases are likely to play pivotal roles in senescence-associated bulk degradation of chloroplast proteins, the exact mechanisms remain a topic of debate [7].

2. Chloroplast recycling

Chloroplasts serve as vital organelles of the plant cell performing photosynthesis as well as production and storage of an array of biomolecules and waste materials. Pigments and light-harvesting proteins make up a significant portion of their nitrogen and Rubisco content. In the advanced stages of aging, chloroplast numbers dwindle. Studies employing electron microscopy have hinted at an intriguing process – chloroplast recycling, the complete self-consumption of chloroplasts, in leaves undergoing senescence induced by aging as well as biotic or abiotic stresses [1].

2.1 Chloroplast recycling in senescence

Senescence, the final stage of leaf development before cell death, includes the disintegration of chloroplast as well as the loss of protein, nucleic acid, pigments, lipids, and polysaccharides (such as starch). Because of this seer degradation, chloroplast ultimately loses either all or a majority of their ability to photosynthesis. The senescence of leaves and other green plant organs results in the annual breakdown of millions of tons of chlorophyll and photosynthetic protein [8]. This results in a loss of photosynthetic capacity, releasing nutrients like nitrogen for re-use in other parts of the plant. During senescence, chloroplast loses their photosynthetic mechanism, and nutrients are redirected to young tissues and storage organs. For instance, nitrogen extracted from mature grains in field-grown rice may represent 60% of the nitrogen released by sensing leaves [9]. In some species, abscission marks the end of senescence, requiring precisely timed petiole senses. As leaves degrade, mesophyll tissue starts to lose its green color and turn yellow, or tissue starts to change color. This is caused by the preferential degradation of chlorophyll over carotenoids, as well as other factors and the synthesis of new compounds such as anthocyanins and phenolic [10]. While chloroplast changes early in senescence, they are last to collapse as the part of this developmental program [11]. Key characterization of chloroplast degeneration includes altered volume and morphology, a transition from ellipsoid to circular shape, and internal membrane reorganization. The three steps that Tamari et al. [12] determined to be the most significant ultrastructural alterations connected with the transformation from mature chloroplast to gerontoplast are thylakoid membrane breakdown, increased plastoglobulus size and number, and modification and disorderness of the plastid envelope. Senescence in chloroplasts in terms of pigment composition resembles with the chromoplasts, yet differs in development, division capacity, retention of the genome, and biosynthetic ability. Chromoplasts have their own DNA, produce carotenoid, are capable of division, and create iron or young chloroplast, but on the other hand, senescent chloroplasts only form from mature chloroplast, are unable to divide, have eliminated all metabolic activity, and do not have their own DNA [13]. The term "gerontoplasts" was introduced to highlight these distinctions. Detecting the exact transition from chloroplast to gerontoplasts is challenging due to difficulties in identifying the senescence onset. During senescence, the natural process of aging and degradation in plant chloroplast undergoes dynamic changes to adapt protein import and export mechanisms to cellular demands. During their light-induced differentiation, chloroplasts import pre-proteins and amino acids from the cytosol. Nuclear-encoded proteins are imported into chloroplast through membrane transport complexes, and protein import is crucial for the regulation of plant adaptation processes. Under adverse stress and environmental conditions including senescence, to ensure regular cell function, the damaged chloroplast or one of its specialized components needs to be dismantled quickly. Similar to this way, during leaf senescence, chloroplast also exports organic materials including proteins and amino acids [14].

2.2 Chloroplast recycling in stress tolerance

Plastids have emerged as pivotal regulators of plant's response to biotic and abiotic stresses. Chloroplasts have the ability to synthesize a variety of secondary metabolites and phytohormones which help out plant cells to withstand adverse conditions. Further, plastids communicate with the nucleus and other cellular organelles for the acquisition of essential molecules for survival under unfavorable conditions.

Chloroplasts act as environmental sensors which not only synthesize molecules for stress tolerance but also induce nucleus-encoded genes to be activated and produce stress-resilient proteins. The MEcPP (methylerythritol cyclodiphosphate) is an important constituent of isoprenoid biosynthesis in plastids and also acts as a retrograde signal to activate the expression of stress-responsive nucleus-encoded genes. MEcPP was first identified by screening for mutants with an elicited expression of HPL (hydroxyperoxide lyase). HPL is a nucleus-encoded stress responsive gene but encodes for a plastid enzyme, which plays a fundamental role in the synthesis of jasmonic acid and other stress-related proteins.

Constitutive expression of HPL led to hyperaccumulation of methylerythritol cyclodiphosphate and regulation of a series of nuclear genes [15]. Likewise, PAP (phosphonucleotide 3′-phosphoadenosine 5′-phosphate) is another important plastid metabolite that has explored retrograde signals for stress tolerance in plants [16]. Plastidic SAL1 (chloroplast 3′ (2′), 5′ -bisphosphate nucleotidase) can dephosphorylate PAP to AMP (adenosine monophosphate). The activity of SAL1 is inhibited, and as a result, PAP accumulation is increased as PAP has been explored to play a crucial role in drought stress tolerance and in the regulation of a number of nucleus-encoded stress-responsive genes [17]. Chloroplasts are sensitive to high-temperature stress, and more than 200 genes have been found to be upregulated in response to heat stress. Plants also induce leaf chlorosis wherein the activity of chlorophyll degrading enzymes is increased during photosynthesis. This rapid degradation of chlorophyll, in response to heat stress, is linked with the activation of genes encoding for pheophytinase and chlorophyllase [18].

Senescence is a key developmental process that has effectively been utilized by the plants to cope with biotic and abiotic stresses. The first organelle to be degraded during senescence is chloroplast which contains more than 75% of the leaf nitrogen. Meanwhile, nucleus and mitochondria remain active and provide energy for the progression of senescence. Most of the nitrogen is in the form of proteins, so proteolysis is the core step for its remobilization particularly during the reproductive stage. The degradation of chloroplast components involves different vesicular pathways which can be dependent and/or independent of the ATG (autophagy-related genes). ATG (autophagy-related gene)-dependent degradation pathways involve RCB (RubisCO-containing bodies) and ATI-PS (ATG8-interacting protein 1-plastid associated bodies). Compared with RubisCO-containing bodies, ATG8-interacting protein does not require functional autophagic machinery. ATI1 interacts with thylakoid localized proteins (PsbS, LHCA4, APE1, and PrxA), whereas RCB transports stroma proteins [19]. Under stress conditions, ATG-dependent pathways help in the removal of damaged chloroplast contents resulting in accelerated senescence and increased sensitivity to abiotic stresses. Hence, these proteins play a crucial role in the adaptation to adverse climatic conditions [20].

Under stress conditions, intensive degradation or remodeling of plastid proteins results in the generation of numerous endogenous peptides which are present in the plant secretome. The stress may lead to an increase in misfolded plastidic proteins which activates the chaperone system for the repair/refolding of damaged proteins as well as induces proteases for the degradation of proteins unable to be refolded. Hence, activity of chloroplast proteolytic machinery plays a crucial role in stress tolerance. Nature has blessed plastids with highly specialized proteases, that is, MAPs (methionine aminopeptidases) for the repair and maturation of not only chloroplast-encoded proteins but also of the nucleus encoded proteins [21]. Hence, plants manage to withstand stresses (biotic and abiotic) through recycling of their chloroplasts which involves degradation of the misfolded proteins as well as the repair/refolding of the damaged proteins [22].

3. Recycling and repair pathways

There are two pathways of plastid recycling and repair, that is, piecemeal degradation where chloroplasts are degraded gradually, while the other pathway is known as chlorophagy in which the whole chloroplast is degraded for recycling.

3.1 Piecemeal degradation of chloroplasts

In the process of piecemeal degradation and chloroplast recycling, a specific mechanism for releasing Rubisco from chloroplasts and subsequently breaking it down in other cell compartments has been proposed as an explanation for the early decline in Rubisco levels compared to the chloroplast population. As previously summarized [23], ATG-dependent autophagy via Rubisco-containing bodies (RCBs), which are sent to the central vacuole for degradation, and an ATG-independent alternative pathway involving senescence-associated vacuoles (SAVs) are currently evident to be at least two distinct transport pathways responsible for breaking down stromal proteins outside the chloroplast.

3.1.1 Rubisco-containing bodies (RCBs)

The RCBs were initially discovered in naturally aging wheat leaves (*T. aestivum*) [4]. According to thorough immunoelectron-microscopic investigations, Rubisco is occasionally contained within tiny spherical structures (RCBs), which are mainly found in the cytoplasm and occasionally in the vacuole. The stromal protein Gln synthetase is found in RCBs, which also have an electron density like that of chloroplast stroma. Notably, RCBs lack major membrane proteins including cytochrome f, LHCII, and the -subunits of ATPase coupling factor 1 as well as thylakoid structures. The isolation membranes (phagophores) that surround the double membranes of RCBs in the cytoplasm are thought to be derived from the chloroplast envelope. In the early phases of leaf senescence, when Rubisco levels begin to decline but chlorophyll content stays largely constant, RCBs are frequently observed.

The phenomenon of gradual chloroplast breakdown through RCBs (Rubisco-containing bodies) seems to be prevalent across various plant species, serving as a crucial mechanism for recycling proteins during both growth and in response to environmental challenges. The RCBs have also been identified in young tobacco leaves (*Nicotiana tabacum*), where they are known as Rubisco-vesicular bodies (RVBs), and in rice leaves (*Oryza sativa*) exposed to salt stress conditions. Interestingly, the RCBs found in rice leaves under salt stress exhibit distinctive structural features compared to previously documented RCBs. These rice RCBs possess inner membrane structures, potentially formed from vesicles originating from invaginations of the chloroplast's inner envelope. Furthermore, they occasionally contain crystalline inclusions that develop within chloroplasts under osmotic stress and disappear during the recovery phase. It is plausible that RCBs play a role in breaking down these crystalline inclusions, and the possible mechanisms involved in RCB formation during salt stress have been extensively detailed [24].

The connection between RCBs (Rubisco-containing bodies) and autophagy has been unveiled in Arabidopsis through the use of reverse genetics and live-cell imaging techniques. In this process, RCBs can be visualized by employing stroma-targeted green fluorescent protein (GFP) (or RFP) or a fusion involving a small subunit of Rubisco (RBCS) with GFP (or RFP). This fusion interacts with the plant's endogenous large Rubisco subunits (RbcL) and RBCS molecules, resulting in the creation of Rubisco-GFP (or RFP) [25]. When Arabidopsis leaves expressing these fluorescent markers are treated with concanamycin A, which suppresses vacuolar lytic activity, spherical structures displaying GFP or RFP fluorescence, devoid of chlorophyll fluorescence, appear within the vacuolar lumen. These structures are referred to as RCBs. Additionally, the accumulation of RCBs is disrupted in autophagy-defective mutants known as atg5 and atg4a4b [25, 26]. In normal, wild-type cells, stroma-targeted RFP and the GFP-ATG8a fusion, serving as a marker for autophagosomes and autophagic bodies, are observed together within autophagic bodies located in the vacuole [27]. These findings provide evidence supporting the notion that RCBs constitute a distinct type of autophagic body containing Rubisco and potentially other proteins localized within the stroma.

3.1.2 Senescence-associated vacuoles (SAVs)

A novel type of diminutive lytic vacuole, known as the senescence-associated vacuole (SAV), has come to light within the cytoplasm of aging leaves in plants like soybean (*Glycine max*), Arabidopsis, and tobacco. This discovery has unveiled an alternative route for the breakdown of chloroplast proteins, distinct from the conventional chloroplast-based degradation [28]. Remarkably, SAVs exclusively emerge in photosynthetic tissues undergoing senescence, harboring a unique senescence-specific cysteine-protease, SAG12 (senescence-associated gene 12).

Comparable to ribulose-1,5-bisphosphate carboxylase/oxygenase (Rubisco) containing bodies (RCBs), SAVs encompass stromal proteins like Rubisco and glutamine synthetase, and stromal-targeted GFP. However, they lack thylakoid proteins such as the LHCII. Additionally, unlike RCBs, SAVs are laden with chlorophyll a. Intriguingly, there is no substantiated evidence thus far to suggest that SAVs engage autophagic machinery. Notably, it was postulated that SAVs still manifest even in the absence of autophagy, as demonstrated in the autophagy-defective atg7 mutant, albeit supporting data were conspicuously absent in the original study [29]. This underscores a semblance in the content of RCBs and SAVs, while their formation mechanisms remain unequivocally distinct. A recent investigation unveiled that SAV buildup coincides with the induction of autophagy in the Arabidopsis des1 mutant, which lacks L-cysteine desulfhydrase 1, an enzyme crucial for cysteine degradation [30]. However, further research is imperative to elucidate the intricate interplay between ATG-dependent autophagy and SAVs.

The visual contrast between SAVs (stroma-associated vacuoles) and RCBs (Rubisco-containing bodies) is quite remarkable and can be easily appreciated under the microscope. The RCBs are normally found enclosed by a double membrane and display internal electron density similar to that of the chloroplast stroma. In sharp contrast, SAVs are enclosed by a single membrane. The electron density within SAVs resembles that of the central vacuole and is notably lower than that found in the chloroplast stroma. Additionally, SAVs often contain dense aggregates within their interior, potentially composed of partially degraded cellular components.

The exact mechanism by which stromal proteins are directed to SAVs remains a mystery. As previously discussed in a comprehensive review [31], one possible pathway involves stromal proteins initially traversing the chloroplast envelope before being transferred directly to SAVs through a yet-to-be-revealed mechanism. Alternatively, similar to the process of piecemeal microautophagy of the nucleus in yeast, it is plausible that SAVs themselves encapsulate a portion of the chloroplast or stromule, forming a structure similar to RCBs. The SAVs exhibit a higher level of acidity compared to the central vacuole and possess potent proteolytic capabilities. Thus, chloroplast particles enclosed within SAVs would undergo rapid degradation, resembling the fate of RCBs within the central vacuole. It is important to note that this phenomenon can only be observed in the presence of concanamycin A.

3.1.3 Protein import/export and chloroplast recycling

The majority of chloroplast proteins originate from nuclear genes and are guided into chloroplast by TOC (translocon at the outer chloroplast membrane) and TIC (translocon at the inner chloroplast membrane) complexes under normal conditions. The process of transporting these nuclear-encoded proteins into the chloroplast is a highly orchestrated series of steps that rely on specific molecules and energy sources. These proteins contain a distinct molecular tag called a transit peptide, which acts as the navigation signal, directing them to the appropriate location within the chloroplast. Several key stages are involved in the import of protein: firstly, they are identified by transit peptide specialized proteins called cytosolic chaperones and targeting factors. These molecules create a protective complex around the protein, preventing it from folding prematurely and maintaining it in a state ready for import. After identification and binding, the protein-chaperon complex intratas with molecular machinery situated at the outer membrane of the chloroplast, referred to as the TOC complex [32]. This nitration triggers the movement of the protein across the outer chloroplast membrane. Upon successfully crossing the outer membrane, the protein chaperon complex encounters another translocon, the TIC complex, embedded in the chloroplast inner membrane [33]. This complex assists in transporting the protein across the inner membrane, ensuring its access to the inner space of the chloroplast. Once the protein has reached the inner space of the chloroplast, the transit peptide is cleaved off. This step is vital for the protein to adopt its final functional shape. Furthermore, molecular chaperones within the chloroplast environment aid in correctly folding the protein, ensuring its effective participation in the intricate biochemical process of organelles. In essence, the process of chloroplast protein import involves a sequence of precisely coordinated events, safeguarding the integrity and functionality of the transported proteins within the dynamic environment. While most proteins are imported, there are situations when specific proteins need to be exported. Some of these exported proteins play roles in photosynthesis, plastid division, or inter-organelle signaling. The export mechanism typically involves recognizing sorting signals and their integration with the export machinery [34].

A controlled physiological process is involved in the gradual breakdown of cellular components leading to significant alterations in chloroplasts' structure and function, thereby influencing the intricate protein import and export process. Prominent observations highlight the changes that take place in such process during senescence. Firstly, studies indicate that the rates of protein import into chloroplast vary as senescence advances. These changes often connect to modifications in the expression of essential components in the TOC and TIC complexes, which play vital roles in facilitating efficient protein translocation across the chloroplast membrane [35]. As chloroplast age during senescence, the importance of quality control mechanisms becomes more prominent. These mechanisms identify and transport mis-folded protein out of the chloroplast to prevent their buildup and potential damage. This involves the retrograde transport of malfunctioning proteins to the cytoplasm for degradation. Senescence introduces distinct sorting signals known as senescence-associated sorting signals. These signals guide proteins to specific locations within the chloroplast, highlighting the unique nature of protein transport during senescence. Moreover, senescence stimulates substantial shifts in gene expression and signal transduction pathways [36]. Regulatory proteins managing this process might need to be exported from chloroplast toward organelles like vacuoles and nucleus. This export plays a vital role in coordinating the different events associated with senescence.

In the process of chloroplast protein degradation, two vesicles' categories play vital roles: Rubisco-containing bodies (RCBs) and senescence-associated vacuoles (SAVs) (**Figure 1**). The RCBs are smaller, sphere-shaped vesicles, and enclosed by dual membranes that consist of Rubisco, excluding thylakoid proteins. They are considered as the autophagosomic bodies responsible for transporting stomata proteins toward the vacuoles [37]. Notably, RCBs operate during darkness, senesce, and leaf carbon homeostasis, participating in Rubisco reduction under stall stress in soybean [38]. In contrast, SAVs are characterized by elevated proteolytic and acidic activity compared to central vacuole. These vessels are prominent during senescence with chloroplast-containing cells, notably accumulating in senescence leaves. They facilitate the breakdown of soluble photosynthesis proteins within the chloroplast stroma, including glutamine synthetase and a large unit of Rubisco. Unlikely autophagy, SAVs can specifically degrade chloroplast components, yet the mechanism of their formation and translocation of chloroplast components to SAVs remains unknown [39].

Recent research has uncovered that SAVs are implicated in the chlorophyll, and PSI degradation throughout senescence, supported by their strong cysteine protease activity, and the presence of the senescence-specific protease perhaps contribute to the breakdown of the Rubisco during leaf senescence [40]. These vesicles are crucial for the proteolysis of chloroplast proteins. Chloroplast proteins need to be moved outside of the plastid for the previously mentioned pathways to function [41]. Lately, a significant gene chloroplast vesiculation (CV) has been recognized, involving translocation of chloroplast proteins pathway external to plastid. This gene's expression has been shown to elevate during ordinary and abiotic stress induced senescence.

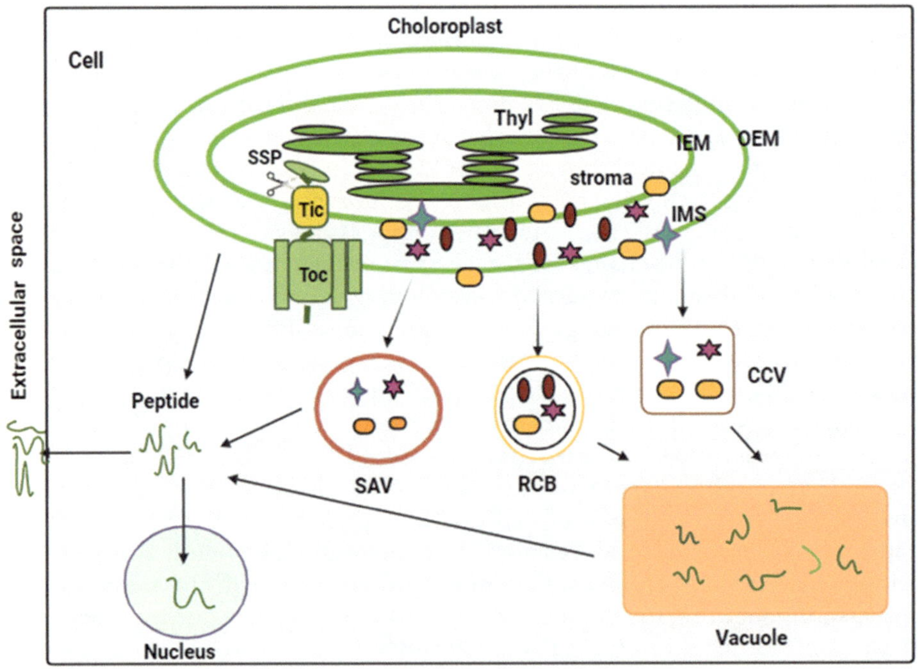

Figure 1.
Cellular pathways involved in chloroplast protein import showing protein trafficking. SAVs represent senescence-associated vacuoles, RCBs are for rubisco-containing bodies, and chloroplast-containing vesicles (CCVs) are for CV-containing vesicles.

Once chloroplast proteins enter the plastids, CV destabilizes the chloroplast, prompting the formation of CCVs during senescence, which subsequently deliver chloroplast proteins to the vacuoles for proteolysis [42]. The CV protein interrelates with several chloroplast proteins, primarily thylakoid components, in plastids, where it localizes. Intra-plastidic vesicles, or CCVs, are produced as a result of the CV gene. These CCVs eventually die from the plastid and travel to the central vacuole, carrying proteins from stromal and thylakoid membranes. Importantly, this degradation process through CCVs operates independently of autophagy and SAVs. The CV protein interacts mainly with thylakoid components, triggering the development of vesicles. The RD26 transcription factor has recently been linked to the induction of the CV-related gene expression. The RD26 regulates the direct degradation of chloroplast proteins during senescence [43]. As chlorophyll breakdown initiates within chloroplasts in senescent leaves, non-fluorescent chlorophyll proteins are released and ultimately broken down in vacuoles. This suggests that senesces or stress can stimulate the degradation of chloroplast protein, even the entire chloroplast, leading to peptide accumulation within the chloroplast, the cytoplasm, and various vesicles. Notably, these peptides could potentially be transported into the extra-cellular space through vesicles and vacuole fusion, aided by outer and inner peptidases, as well as peptide and amino acid transporters.

3.2 Role of autophagy in chloroplast recycling

Autophagy is a highly conserved biological mechanism in eukaryotic organisms. This process involves the engulfing of cellular components within membrane-bound vesicles, followed by their subsequent degradation through lytic processes. Whole organelles can also be enclosed within these membrane structures as part of autophagy [44]. These cellular components wrapped in membrane-bound vesicles are taken to special compartments called vacuoles or lysosomes. Inside these compartments, enzymes called hydrolases break down both the vesicles and their contents. There are two different forms of autophagy, known as micro-autophagy and macro-autophagy, which have been seen in various organisms, including plants [45]. In micro-autophagy, parts of the cell's fluid are directly taken in by the folding of the vacuolar membrane. In the more common form called macro-autophagy, often just referred to as autophagy, the cell's content is enclosed within a double-layered structure called an autophagosome. The fusion event between the outer membrane of the autophagosome and the vacuolar membrane facilitates the transfer of the enclosed inner membrane-defined structure, known as the autophagic body, into the interior space of the vacuole.

In more recent times, a breakthrough occurred with genome-wide studies, which paved the way for a molecular examination of plant autophagy. These studies identified numerous genes in plants that are similar to yeast ATGs and also explored the effects of disrupting these genes in Arabidopsis (*Arabidopsis thaliana*). Researchers have explored the roles of Arabidopsis ATGs by utilizing T-DNA insertional knock-out mutants and RNA interference knockdown mutants. Additionally, they have established a live-cell system for monitoring autophagy in plants. This system involves the use of a GFP fused with ATG8, which acts as a marker to track the presence of autophagosomes. These molecular methods have demonstrated the crucial function of plant autophagy in response to food deficiency, abiotic stressors, and pathogen infection. They have also confirmed that the central autophagy apparatus functions in plants but not in yeast [46]. Additionally, it has been demonstrated that target of rapamycin (TOR) kinase functions as a negative regulator in yeast and Arabidopsis [47].

Plant autophagy was initially studied for its ability to respond to nutrient deprivation, a critical role similar to that of yeast. Autophagy is triggered when cultured plant cells lack externally supplied sucrose. During this period, key ATGs linked to Ub-like conjugation systems experience temporary increases in activity [48]. Scientists have examined the role of autophagy in nutrient recycling by studying autophagy-deficient (atg) mutant plants. Essentially, Arabidopsis atg mutants can complete their life cycles [1].

However, they struggle to survive extended periods of nitrogen and/or carbon starvation. Additionally, they exhibit accelerated leaf aging and cell death, along with reductions in chlorophyll and photosynthetic proteins, even when nutrient and growth conditions are favorable. This premature aging affects seed production, leading to lower yields in atg mutants. Consequently, it was previously concluded that while autophagy plays a vital role in nutrient recycling during starvation and senescence in plants, it does not primarily target chloroplasts, despite their abundance in leaf degradation [45].

3.3 Chlorophagy: autophagy of whole chloroplasts

In the advanced stages of aging, chloroplast numbers dwindle. Studies employing electron microscopy have hinted at an intriguing process – the complete self-consumption of chloroplasts, an occurrence termed "chlorophagy," in leaves undergoing senescence induced by darkness [1]. In the world of Arabidopsis, a plant's individual leaf darkened for experimentation experiences a rapid onset of senescence. Remarkably, within a few short days, both the quantity and size of chloroplasts within these leaves diminish significantly. However, in the case of individually darkened leaves (IDLS) from the atg4s mutant, senescence, evident as chlorosis, unfolds much like in the wild-type, yet the drop in chloroplast count and to some extent, their size, is hindered [49].

What's more, beyond the realm of conventional chloroplast behavior, small chloroplasts retaining their chlorophyll fluorescence come into view inside the vacuole after 3 days of the IDL treatment in wild-type specimens, but this phenomenon remains absent in the atg4 mutant. Given that the formation of Rubisco-containing bodies (RCBs) consumes elements from both the stroma and the chloroplast envelope, it is a logical assumption that this process contributes to the shrinking of chloroplasts. These shrunken chloroplasts, reminiscent of gerontoplasts, are then ferried into the vacuole through the machinery of autophagy.

Notably, mutations that disrupt the normal functioning of chloroplasts can trigger the wholesale degradation of these organelles, a process likely attributable to chlorophagy. This degradation appears to occur independently of senescence or starvation. For instance, partially degraded chloroplasts make their home within the vacuoles of cotyledon cells belonging to the Arabidopsis ppi40 mutant, a genetic variant deficient in a protein homologous to the 40 kDa protein found in the inner envelope membrane of chloroplasts' translocon (Tic complex). In the advanced stages of aging, chloroplast numbers dwindle. Studies employing electron microscopy have hinted at an intriguing process – the complete self-consumption of chloroplasts, an occurrence termed "chlorophagy," in leaves undergoing senescence induced by darkness [1]. As chloroplast transfer to the vacuole is observed even under well-fed conditions, it is posited that plants utilize autophagy to eliminate defective plastids, ensuring quality control of their organelles.

The Arabidopsis mex1 mutant, which lacks the maltose transporter in the chloroplast envelope, presents a fascinating scenario. This mutant accumulates higher-than-normal levels of maltose and starch within its chloroplasts and displays a chlorotic phenotype even when it's not undergoing senescence. What's particularly intriguing is that various components of chloroplasts, such as thylakoid membranes, starch granules, and plastoglobules, have been found inside the vacuole of the mex1 mutant [50]. These discoveries have given rise to a hypothesis suggesting that the increased maltose concentration in the mex1 mutant disrupts chloroplast function, possibly setting off a form of retrograde signaling that initiates the degradation of chloroplasts.

4. Conclusions

Green photosynthetic organisms including plants, algae, and cyanobacteria are the source of energy and carbon fixation through photosynthesis. Green pigments containing plastid "chloroplasts" are the hub of photosynthesis. It supplies energy to all living organisms. Photosynthesis products, through various biochemical pathways, produce numerous macromolecules. Moreover, chloroplasts become an abundant source of necessary organic nitrogen and other nutrients through the process of chloroplast recycling in senescing leaves. Such nutrients are utilized in growth, reproduction, storage organs, and new plant organs. The protein translocation mechanism is essential for chloroplast function and plays a critical role in cellular adaptations during senescence. The dynamic changes observed in these processes during senescence highlight their significance in plant adaptation to changing environmental conditions and developmental stages. Though limited literature is available, a deep understanding of the protein import/export during senescence and/or chloroplast recycling can potentially lead to the progression of strategies to enhance crop productivity and stress tolerance, contributing to agriculture sustainability and food security.

Author details

Faiz Ahmad Joyia, Ghulam Mustafa and Muhammad Sarwar Khan*
Faculty of Agriculture, Center of Agricultural Biochemistry and Biotechnology (CABB), University of Agriculture, Faisalabad, Pakistan

*Address all correspondence to: sarwarkahn_40@hotmil.com

IntechOpen

© 2024 The Author(s). Licensee IntechOpen. This chapter is distributed under the terms of the Creative Commons Attribution License (http://creativecommons.org/licenses/by/3.0), which permits unrestricted use, distribution, and reproduction in any medium, provided the original work is properly cited.

References

[1] Ishida H, Masanori I, Shinya W, Amane M. Roles of autophagy in chloroplast recycling. Biochimica et Biophysica Acta. 2014;**1837**:512-521

[2] Araujo WL, Tohge T, Ishizaki K, Leaver CJ, Fernie AR. Protein degradation — An alternative respiratory substrate for stressed plants. Trends in Plant Science. 2011;**16**:489-498

[3] Martinez DE, Bartoli CG, Grbic V, Guiamet JJ. Vacuolar cysteine proteases of wheat (*Triticum aestivum* L.) are common to leaf senescence induced by different factors. Journal of Experimental Botany. 2007;**58**:1099-1107

[4] Chiba A, Ishida H, Nishizawa NK, Makino A, Mae T. Exclusion of ribulose-1,5-bisphosphate carboxylase/oxygenase from chloroplasts by specific bodies in naturally senescing leaves of wheat. Plant & Cell Physiology. 2003;**44**:914-921

[5] Kato Y, Murakami S, Yamamoto Y, Chatani H, Kondo Y, Nakano T, et al. The DNA-binding protease, CND41, and the degradation of ribulose-1, 5-bisphosphate carboxylase/oxygenase in senescent leaves of tobacco. Planta. 2004;**220**:97-104

[6] Sakamoto W. Protein degradation machineries in plastids. Annual Review of Plant Biology. 2006;**57**:599-621

[7] Wagner R, Aigner H, Pruzinska A, Jankanpaa HJ, Jansson S, Funk C. Fitness analyses of *Arabidopsis thaliana* mutants depleted of FtsH metalloproteases and characterization of three FtsH6 deletion mutants exposed to high light stress, senescence and chilling. The New Phytologist. 2011;**191**:449-458

[8] Gregersen PL, Culetic A, Boschian L, Krupinska K. P. Senescence and crop productivity. Plant Molecular Biology. 2013;**82**:603-622

[9] Ramkumar MK, Senthil Kumar S, Gaikwad K, Pandey R, Chinnusamy V, Singh NK, et al. A novel stay-green mutant of rice with delayed leaf senescence and better harvest index confers drought tolerance. Plants. 2019;**8**(10):375

[10] Wang J, Huang R. Modulation of ethylene and ascorbic acid on reactive oxygen species scavenging in plant salt response. Frontiers in Plant Science. 2019;**10**:319

[11] Hörtensteiner S, Krautler B. Chlorophyll breakdown in higher plants. Biochimica et Biophysica Acta - Bioenergetics. 2011;**1807**(8):977-988

[12] Tamary E, Nevo R, Naveh L, Levin-Zaidman S, Kiss V, Savidor A, et al. Chlorophyll catabolism precedes changes in chloroplast structure and proteome during leaf senescence. Plant Direct. 2019;**3**(3):1-18

[13] Thomas H, Ougham H, Canter P, Donnison I. What stay-green mutants tell us about nitrogen remobilization in leaf senescence. Journal of Experimental Botany. 2002;**53**(370):801-808

[14] Buet A, Costa ML, Martínez DE, Guiamet JJ. Chloroplast protein degradation in senescing leaves: Proteases and lytic compartments. Frontiers in Plant Science. 2019;**10**:747

[15] Lemos M, Xiao Y, Bjornson M, Wang JZ, Hicks D, Souza A, et al. The plastidial retrograde signal methyl erythritol cyclopyrophosphate is a regulator of salicylic acid and jasmonic acid crosstalk. Journal of Experimental Botany. 2016;**67**:1557-1566

[16] Chan KX, Mabbitt PD, Phua SY, Mueller JW, Nisar N, Gigolashvili T, et al. Sensing and signaling of oxidative stress in chloroplasts by inactivation of the SAL1 phosphoadenosine phosphatase. Proceedings of the National Academy of Sciences of the United States of America. 2016;**113**:E4567-E4576

[17] Song Y, Feng L, Alyafei MAM, Jaleel A, Ren M. Function of chloroplasts in plant stress responses. International Journal of Molecular Sciences. 2021;**22**:13464. DOI: 10.3390/ijms222413464

[18] Jiao B, Meng Q, Lv W. Roles of stay-green (SGR) homologs during chlorophyll degradation in green plants. Botanical Studies. 2020;**61**:25

[19] Zhuang X, Jiang L. Chloroplast degradation: Multiple routes into the vacuole. Frontiers in Plant Science. 2019;**10**:1-8. DOI: 10.3389/fpls.2019.0035

[20] Xie Q, Michaeli S, Peled-Zehavi H, Galili G. Chloroplast degradation: One organelle, multiple degradation pathways. Trends in Plant Science. 2015;**20**:264-265. DOI: 10.1016/j.tplants.2015.03.013

[21] Ghifari AS, Huang S, Murcha MW. The peptidases involved in plant mitochondrial protein import. Journal of Experimental Botany. 2019;**21**:6005-6018

[22] Nishimura K, Kato Y, Sakamoto W. Essentials of proteolytic machineries in chloroplasts. Molecular Plant. 2017;**10**:4-19

[23] Reumann S, Voitsekhovskaja O, Lillo C. From signal transduction to autophagy of plant cell organelles: Lessons from yeast and mammals and plant-specific features. Protoplasma. 2010;**247**(247):233-256

[24] Yamane K, Mitsuya S, Taniguchi M, Miyake H. Salt-induced chloroplast protrusion is the process of exclusion of ribulose-1,5-bisphosphate carboxylase/oxygenase from chloroplasts into cytoplasm in leaves of rice. Plant, Cell & Environment. 2012;**35**:1663-1671

[25] Ishida H, Yoshimoto K, Izumi M, Reisen D, Yano Y, Makino A, et al. Mobilization of rubisco and stroma-localized fluorescent proteins of chloroplasts to the vacuole by an ATG gene-dependent autophagic process. Plant Physiology. 2008;**148**:142-155

[26] Wada S, Ishida H, Izumi M, Yoshimoto K, Ohsumi Y, Mae T, et al. Autophagy plays a role in chloroplast degradation during senescence in individually darkened leaves. Plant Physiology. 2009;**149**:885-893

[27] Izumi M, Wada S, Makino A, Ishida H. The autophagic degradation of chloroplasts via rubisco-containing bodies is specifically linked to leaf carbon status but not nitrogen status in Arabidopsis. Plant Physiology. 2010;**154**:1196-1209

[28] Martinez DE, Costa ML, Gomez FM, Otegui MS, Guiamet JJ. 'Senescence-associated vacuoles' are involved in the degradation of chloroplast proteins in tobacco leaves. The Plant Journal. 2008;**56**:196-206

[29] Otegui MS, Noh YS, Martinez DE, Petroff MGV, Staehelin LA, Amasino RM, et al. Senescence-associated vacuoles with intense proteolytic activity develop in leaves of Arabidopsis and soybean. The Plant Journal. 2005;**41**:831-844

[30] Alvarez C, Garcia I, Moreno I, Perez-Perez ME, Crespo JL, Romero LC, et al. Cysteine-generated sulfide in the cytosol negatively regulates autophagy

and modulates the transcriptional profile in Arabidopsis. The Plant Cell. 2012;**24**:4621-4634

[31] Martinez DE, Costa ML, Guiamet JJ. Senescence-associated degradation of chloroplast proteins inside and outside the organelle. Plant Biology. 2008;**10**:15-22

[32] Chiu CC, Chen LJ, Li HM. Pea chloroplast DnaJ-J8 and Toc12 are encoded by the same gene and localized in the stroma. Plant Physiology. 2010;**154**(3):1172-1182

[33] Gutensohn M, Fan E, Frielingsdorf S, Hanner P, Hou B, Hust B, et al. Toc, Tic, Tat et al.: Structure and function of protein transport machineries in chloroplasts. Journal of Plant Physiology. 2006;**163**(3):333-347

[34] Lee DW, Lee S, Oh YJ, Hwang I. Multiple sequence motifs in the rubisco small subunit transit peptide independently contribute to Toc159-dependent import of proteins into chloroplasts. Plant Physiology. 2009;**151**(1):129-141

[35] Qbadou S, Becker T, Bionda T, Reger K, Ruprecht M, Soll J, et al. Toc64 - A preprotein-receptor at the outer membrane with bipartide function. Journal of Molecular Biology. 2007;**367**(5):1330-1346

[36] Guiboileau A, Avila-Ospina L, Yoshimoto K, Soulay F, Azzopardi M, Marmagne A, et al. Physiological and metabolic consequences of autophagy deficiency for the management of nitrogen and protein resources in Arabidopsis leaves depending on nitrate availability. New Phytologist. 2013;**199**(3):683-694

[37] Izumi M, Nakamura S. Partial or entire: Distinct responses of two types of chloroplast autophagy. Plant Signaling & Behavior. 2017;**12**(11):e1393137

[38] He Y, Yu C, Zhou L, Chen Y, Liu A, Jin J, et al. Rubisco decrease is involved in chloroplast protrusion and rubisco-containing body formation in soybean (Glycine max) under salt stress. Plant Physiology and Biochemistry. 2014;**74**:118-124

[39] Carrión CA, Costa ML, Martínez DE, Mohr C, Humbeck K, Guiamet JJ. In vivo inhibition of cysteine proteases provides evidence for the involvement of "senescence-associated vacuoles" in chloroplast protein degradation during dark-induced senescence of tobacco leaves. Journal of Experimental Botany. 2013;**64**(16):4967-4980

[40] Gomez FM, Carrión CA, Costa ML, Desel C, Kieselbach T, Funk C, et al. Extra-plastidial degradation of chlorophyll and photosystem I in tobacco leaves involving 'senescence-associated vacuoles'. Plant Journal. 2019;**99**(3):465-477

[41] James M, Poret M, Masclaux-Daubresse C, Marmagne A, Coquet L, Jouenne T, et al. SAG12, a major cysteine protease involved in nitrogen allocation during senescence for seed production in Arabidopsis thaliana. Plant and Cell Physiology. 2018;**59**(10):2052-2063

[42] Wang S, Blumwald E. Stress-induced chloroplast degradation in arabidopsis is regulated via a process independent of autophagy and senescence-associated vacuoles. Plant Cell. 2019;**26**(12):4875-4888

[43] Kamranfar I, Xue GP, Tohge T, Sedaghatmehr M, Fernie AR, Balazadeh S, et al. Transcription factor RD26 is a key regulator of metabolic reprogramming during dark-induced

senescence. New Phytologist. 2018;**218**(4):1543-1557

[44] Nakatogawa H, Suzuki K, Kamada Y, Ohsumi Y. Dynamics and diversity in autophagy mechanisms: Lessons from yeast. Nature Reviews. Molecular Cell Biology. 2009;**10**:458-467

[45] Bassham DC, Laporte M, Marty F, Moriyasu Y, Ohsumi Y, Olsen LJ, et al. Autophagy in development and stress responses of plants. Autophagy. 2006;**2**:2-11

[46] Suttangkakul A, Li F, Chung T, Vierstra RD. The ATG1/ATG13 protein kinase complex is both a regulator and a target of autophagic recycling in Arabidopsis. The Plant Cell. 2011;**23**:3761-3779

[47] Liu Y, Bassham DC. TOR is a negative regulator of autophagy in Arabidopsis thaliana. PLoS One. 2010;**5**:e11883

[48] Rose TL, Bonneau L, Der C, Marty-Mazars D, Marty F. Starvation-induced expression of autophagy-related genes in Arabidopsis. Biology of the Cell. 2006;**98**:53-67

[49] Keech O, Pesquet E, Ahad A, Askne A, Nordvall D, Vodnala SM, et al. The different fates of mitochondria and chloroplasts during dark-induced senescence in Arabidopsis leaves. Plant, Cell & Environment. 2007;**30**:1523-1534

[50] Stettler M, Eicke S, Mettler T, Messerli G, Hortensteiner S, Zeeman SC. Blocking the metabolism of starch breakdown products in Arabidopsis leaves triggers chloroplast degradation. Molecular Plant. 2009;**2**:1233-1246

Chapter 6

The Role of Guard Cells Chloroplasts toward the Enhancement of Plant Growth

Batta Kucheli

Abstract

Guard cells contain chloroplasts, and the stomata through which exchange of gas takes place. They control the stomatal pore, which serves as a channel for exchange of gas by balancing between CO_2 uptake for photosynthesis and water loss through transpiration. As a result, chloroplasts in the guard cells have become potential tool for manipulation toward improvement of plant productivity through photosynthesis. The role of the guard cells chloroplasts can, therefore, be elucidated through manipulations of enzymes for photosynthesis by using molecular means. The cytochrome b_6f complex catalyzes the transfer of electrons between the two photosynthetic reaction centers, Photosystems II and Photosystem I, while at the same time, transferring protons across the thylakoid used to synthesize ATP for the Calvin cycle. In this study, the overexpression of the Rieske FeS protein in Arabidopsis exhibited phenotypes, which resulted in substantial improvements of quantum efficiency of PSII. Transgenic lines were significantly higher in early development of the plants. Phenotypes observed in the transformed plants also showed faster initial growth rates evidenced by larger leaf area and faster rosette increases, which may suggest that Rieske might be of importance for enhanced plant growth. The result obtained proves more opportunities await the exploitation of guard cells chloroplasts metabolism toward the improvement of plants.

Keywords: Rieske FeS, guard cells, photosynthesis, chloroplasts, plant growth

1. Introduction

Stomatal conductance determines the flux of gases between the inside of the leaf and the external atmosphere, which influences photosynthetic carbon assimilation and water use. Understanding the structure, function, and signaling mechanism in stomata in response to changing environmental conditions is, therefore, critically important if we are to manipulate the processes for optimal plant use (**Figure 1**) [1–3].

Stomata has, therefore, attracted the attention of scientists for almost three centuries [4], and a great deal of knowledge related to the structure, development, and physiology of stomata has been acquired. In order to balance between carbon supply and the ability to optimize or sustain plant growth in an ever-fluctuating environment, an understanding of the external and internal responses is required, as

 IntechOpen

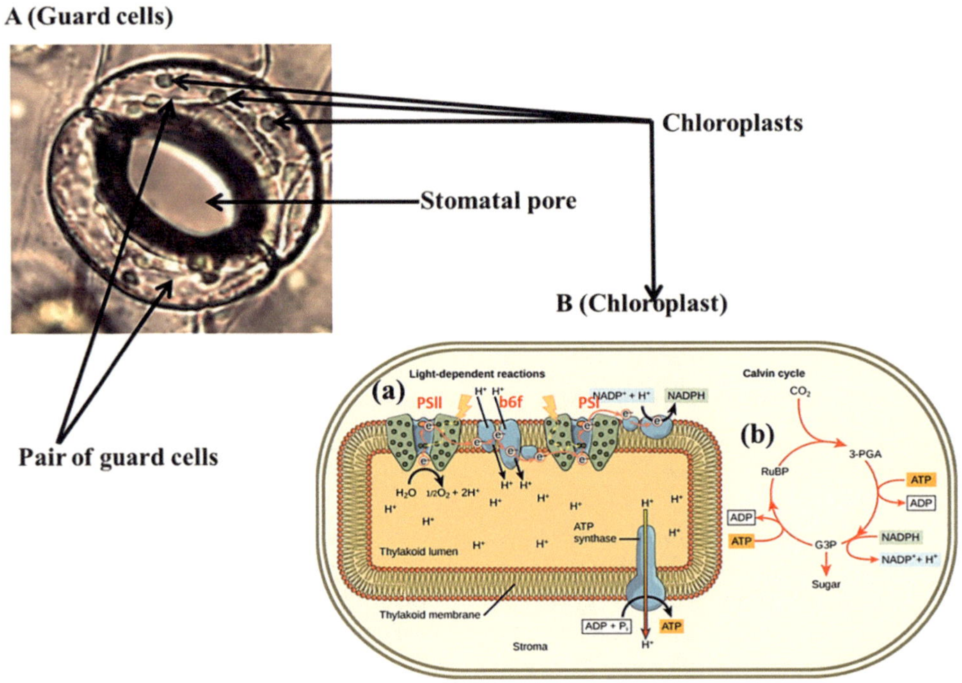

Figure 1.
Guard cells, stomata, and chloroplasts as potential tools for manipulation. (A) Guard cells from an epidermal peel showing chloroplasts, which are the site of photosynthesis and the stomata through which exchange of gas takes place. The guard cells control the stomatal pore, that is. The opening and closing of the stomata. (B) Schematic diagram of a single chloroplast illustrating (a) the thylakoid membrane where electron transport chain takes place. The cytochrome b_6f between the PSII and PSI shuttles electrons, which are used for the synthesis of NADPH and ATP as energy for fueling the Calvin cycle. (b) Calvin cycle found in the stroma of the chloroplast where sedoheptulose-1, 7-bisphosphatase (SBPase) a key component in the regeneration of RuBP in the Calvin cycle functions. (source: Picture taken from PhD research of Batta kucheli 2018).

well as new approaches that integrate both molecular and physiological approaches [5, 6]. Genetic potentials for maximum yield still lie unrealized, and the need for better adaptation of plants to climatic factors in which the plants are grown are still enormous [7].

In view of this, this chapter highlights or provides a way to elucidate the role of guard cell chloroplasts by using molecular tools and techniques to manipulate chloroplast metabolism specifically in the guard cells.

1.1 Guard cell chloroplasts

Photosynthesis takes place primarily in the mesophyll tissue as epidermal cells generally lack chloroplasts. However, guard cells, which developed from protodermal cells, also contain photosynthetically active chloroplasts in most but not all species [8–12]. Guard cell chloroplasts are smaller and have less granum, hence could be said to be less developed than the mesophyll cells. Guard cell chloroplasts have a reduced thylakoid network and chlorophyll contents compared to the mesophyll [13] and have functional photosystems I and II, electron transport, oxygen evolution, and photophosphorylation [8, 14]. Elucidating the role of the guard cell chloroplasts and guard cell photosynthesis presents a serious challenge for researchers in the field but the following roles have been proposed for guard cell chloroplasts.

1. Electron transport in guard cells is capable of producing ATP used in osmoregulation [15–17].

2. Photosynthetic carbon assimilation within guard cells produces osmotically active sugars [8, 12, 17, 18].

3. Chloroplasts are involved in blue-light signaling and response [19, 20].

4. Starch stored in the chloroplasts (either produced from carbon assimilation in the guard cell chloroplasts or imported from the mesophyll) is available to synthesize malate as a counter ion to K^+ or is degraded into sucrose.

1.2 Electron transport in guard cells

Linear electron transport and photophosphorylation in the guard cells have been documented. The quantum efficiency for PSII photochemistry in guard cells has been shown to the rate of 70 to 80% that of mesophyll cells when subjected to a range of light levels thereby suggesting a similarity of operation in mechanisms in both guard and mesophyll cell [14, 21]. The pigment composition of guard cells is also similar to that of the mesophyll cells [22], which leads to enhance ATP production. As a result, such electron transport rate could provide sufficient energy to drive ions for stomatal opening in the absence of CO_2 fixation. Using the high-resolution chlorophyll fluorescence imaging, [14, 23] found that Rubisco was a major sink for the products of electron transport suggesting that guard cell electron transport can be mediated by [CO_2] meaning Calvin cycle activity does take place in the guard cell [24].

1.3 Advances, tools, and techniques for chloroplasts and guard cells manipulation toward plants improvements

Recent developments in technology have opened opportunities to explore guard cell mechanisms, and therefore potentials for regulating WUE and enhancing plant productivity. Efficient, simple, and fast cloning techniques are available for the design of desired single or multiple genes to be expressed in plants for genetic manipulation of metabolites involved in photosynthetic metabolism, which could lead to changes in stomatal behavior and potentially improve photosynthesis and water use efficiency in plants [25].

Specific cell metabolisms in guard and mesophyll cells can now be targeted and the possible coordination between mesophyll metabolites in relation to stomatal functions determined. The development of guard cell-specific promoters has made these manipulations of expression of specific gene transcripts possible, which provides opportunities to manipulate guard cell-specific metabolisms or specific stomatal traits in order to elucidate their functions or interactions. For instance, KST1 and MYB60 promoters, which are guard cell-specific promoters developed by Müller-Röber et al. [26], have been used to drive the expression of target genes specifically in guard cells. The use of a specific guard cell promoter has demonstrated by Wang et al. [27], Kucheli [25] through an enhanced light-induced stomatal opening, greater photosynthesis, and improved growth rate in Arabidopsis overexpressing H^+-ATPase among others.

Prior studies have targeted both the Calvin cycle and electron transport chain enzymes to expressions in order to identify their control on carbon assimilation [28–30]. It is now possible to demonstrate single or more enzyme transformations, which have shown to enhance photosynthetic rates in varieties of crops. For instance, increased sedoheptulose-1,7-bisphosphatase activity resulted in tobacco plants with improvements in carbon assimilation by 6–12% [28, 31]. It is in line with this that we elucidated the role of the guard cells chloroplasts using guard cell-specific promoters (Myb60 and KST1) to drive expression specifically in the guard cells in order to produce transgenic plants with manipulation in electron transport chain.

1.4 Design and development of golden gate construct to manipulate expression of cytochrome B_6F (Rieske) in *Arabidopsis thaliana*

An electron transport chain is a series of complex reactions that transfer electrons from electron donors to electron acceptors alongside the transfer of protons H^+ ions across a membrane. The cytochrome b_6f complex also known as the plastoquinol-plastocyanin reductase is an enzyme found in the electron transport chain on the thylakoid membrane of chloroplasts. In photosynthesis, the cytochrome b_6f complex catalyzes the transfer of electrons between the two photosynthetic reaction centers, Photosystems II and Photosystem I, while at the same time, transferring protons across the thylakoid used to synthesize ATP from ADP for the Calvin cycle [32, 33].

The overexpression of the Rieske FeS protein in Arabidopsis resulted in substantial improvements of quantum efficiency of PSI and PSII and electron transport, which lead to significant impacts on plant yield [30].

Transgenic approaches have demonstrated striking results of the manipulation of the Calvin cycle where energy conversion led to increasing yield potential [29, 34–37] suggesting that these studies may imply improvements in photosynthesis through overexpressing the activity of individual enzymes. Already, evidence supporting this hypothesis from single manipulations has been demonstrated from transgenic tobacco plants over-expressing SBPase [28] and also the combined multigene approach of over-expressing SBPase and FBPA [38]. These photosynthetic manipulations resulted in increased carbon assimilation, enhanced growth, and increased cumulative biomass, hence the genetic potentials that lie there. It is, therefore, obvious that number of sense and antisense plants with increased and reduced levels of SBPase and Rieske have varying photosynthetic capacity and have altered carbohydrate status at the whole leaf level, thus leading to modifications in growth and development.

Transgenic plants with guard cells specific manipulation are, therefore, important for identification and characterization of the signal transduction mechanisms [25]. Therefore, guard cell photosynthesis demands genetic analyses by guard cell-specific manipulation of photosynthesis. Multiple targets have been identified and could be manipulated to aid more understanding to maximize crop production. Some of these targets are the SBPase and Rieske enzymes, which have demonstrated of their significance in controlling photosynthetic processes.

2. Materials and methods

2.1 Design and construction of constructs for use in *Arabidopsis thaliana* in the chloroplasts of guard cells

The sequences of the genes of interest Rieske in *Arabidopsis thaliana* were retrieved from the TAIR database and primers designed. The plasmid vectors used for the plant transformation were constructed using the golden gate cloning and the Moclo system [39, 40]. Constructs were designed to alter expression of the Rieske genes in a cell-specific manner driven by the KST1 and MYB60 promoters. YFP tags were also included in the construct to demonstrate cell specificity.

2.2 Screening of mutants

Arabidopsis mutants were identified by PCR screening. Arabidopsis stable transformants carrying the transgenes were also screened using antibiotic and/or herbicide. Plants were germinated on agar plates containing antibiotics. For seeds germinated in soil, the soil was treated with 0.82 mM of glufosinate-ammonium and were watered with this until selection was obvious and seedling selected with plants transplanted into individual pots. The presence of the transgene was reconfirmed by genomic DNA PCR screening.

3. Results

3.1 Selection of *Arabidopsis* transformants

Transformed floral plants in the Arabidopsis plants were allowed to mature to seed. Seeds were collected and planted for the screening of the T1 generation and selection of positive transformants achieved. Selection of positive transformants was identified by the application of BASTA watered on the soil in which the T1 germinated seedlings as transformants were resistant to the herbicide (**Figure 2**).

The resulting successful transgenic plants were selected on the herbicide glofusinate ammonium (BASTA) and subsequent screening for homozygous lines began by PCR followed by confirmation by iDNA technology.

4. DNA analysis of transformed plants

The result selected from transformed individual plants was checked for the presence of the transgene by PCR analysis. The result of the DNA analysis produced PCR fragment sizes exactly as the gene of interest in all the lines screened. Constructs have shown all ten lines selected positive for the presence of the transgene (**Figure 3**).

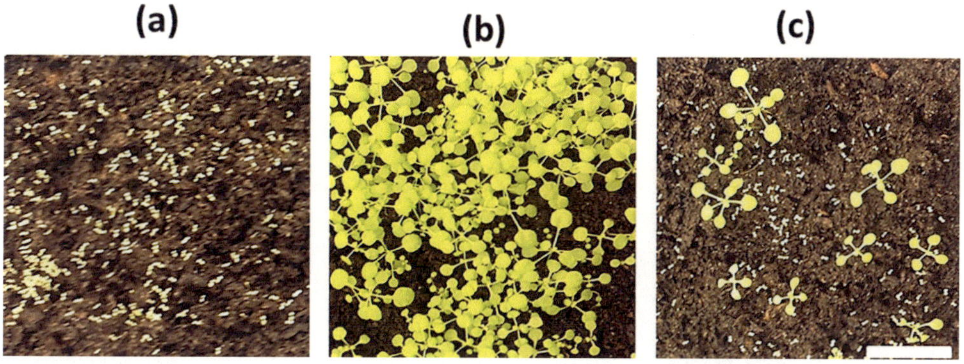

Figure 2.
Hebicide (BASTA) selection of transformed Arabidopsis plants. Resistant transformants were selected by growing on soil and spraying with BASTA (presence of bar gene confers resistance to the glofusinate ammonium herbicide BASTA). (a) WT control showing complete death of cotyledons grown on BASTA, (b) WT control grown without BASTA, and (c) selection of resistant transformants grown on soil watered with BASTA. Plants growth conditions maintained under controlled-environment growth room with (22°C, 8 h light, 16 h dark cycle). White scale bar represents 1 cm.

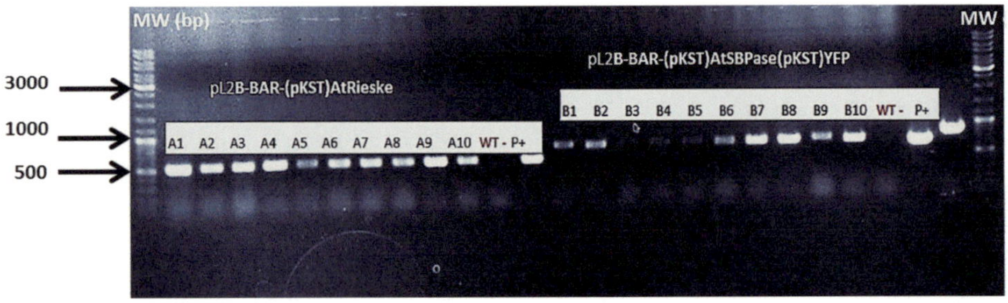

Figure 3.
Genomic DNA PCR screening of transformants for presence of the transgene. The presence of transgenes was checked by PCR analysis of genomic DNA of plants). Ten lines were screened per construct. WT DNA (WT + red) and plasmid DNA containing the gene of interest (P+) were used as negative and positive controls, respectively. PCR products were run alongside molecular weight marker (DNA generuler ladder mix from thermoscientific) in base pairs.

5. Fluorescence microscopy to detect YFP expression in chloroplasts of guard cells

Rieske YFP mutants were rapidly screened for the presence and localization of the yellow fluorescence protein (YFP) specifically in guard cells using high-resolution chlorophyll fluorescence microscope. Constructs fused to the yellow fluorescence protein confirmed that expressions driven by the cells specific promoters were confined to the chloroplasts in the guard cells. Constructs tagged with either the MYB60 promoter or KST1 promoter revealed the chloroplasts with the YFP in them while wild-type control had no signal (**Figure 4**).

6. Photosystem II operating efficiency and growth analysis

The first obvious thing observed in the transgenic lines was the clear evidence of phenotypes in all the constructs selected for further analysis.

Observations revealed developmental phenotypes in the early stage of plants growth between the WT and transgenic lines. The total rosettes or leaf area of the

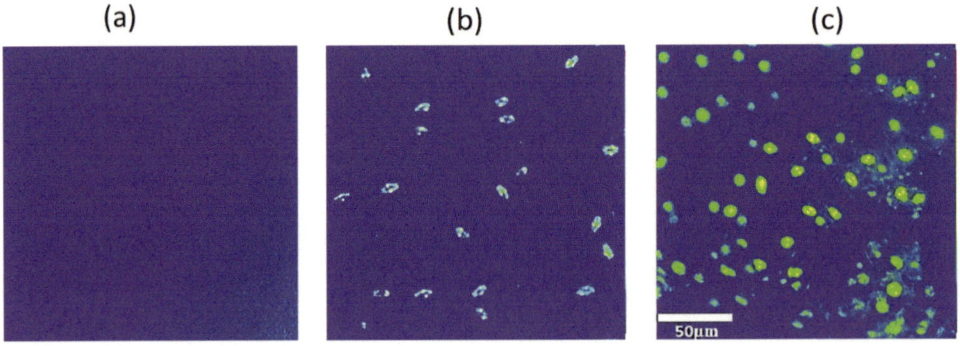

Figure 4.
Specific YFP expression in the guard cell of transformed plants. Localization of the YFP in the chloroplasts of guard cells of Arabidopsis transformants. (a) Wild-type (Col-0) tissue showing no signal while (b) expression of the constructs tagged with YFP and driven by the MYB60 promoter in leaf tissue was checked using the high-resolution microscope. Images acquired by exiting with 515 nm LEDs and emission collected with a band pass filter 530 ± 20).

transgenic plant of the lines within the construct 4-pL2B-BAR-(pKST1)-AtRieske-tHSP evidently showed larger leaf area (**Figure 5**).

Chlorophyll fluorescence imaging was also performed on the plants and the operating efficiecny of PSII photochemistry (*Fq'/Fm'*) determined. Chlorophyll fluorescence imaging of WT and mutant lines of construct 4-(pL2B-BAR-(pKST1)-AtRieske-tHSP to changes in light intensity subjected to 150 and 600 μmol m $^{-2}$ s^{-1} showed significant differences between the wild type and mutants in the first 3 and 4 weeks of recording confirming also the phenotype found in the images observed above. However, as the plants advanced in the next 2–3 weeks, the reductions in *Fq'/Fm'* of the wild type seemed to catch up with the transgenic lines in all constructs suggesting that the expression of these genes might be most critical in their early stages of development. There were no significant differences found in the later weeks of the experiment between the wild type and the transgenics (**Figure 6**).

Chlorophyll fluorescence imaging used to determine the maximum PSII operating efficiency of photosynthesis (*Fq'/Fm'*) values of the whole plant was also used to capture images at the time of analysis. These images further tell more of the PSII efficiency differences indicated by colors from the scale, which ranges from green (lowest value) to orange (highest value). The more the value, the more efficient is the PSII efficiency. Images likewise show the growth differences, which were evident between the wild type and transgenic lines. The PSll photosynthetic differences of all the four constructs at week 3 are shown in the **Figure 7** below, which clearly

Figure 5.
Growth phenotype of WT and homozygous mutant lines of construct 4- (pL2B-BAR-(pKST1)-AtRieske-tHSP plants grown on soil. 4–weeks old plants were germinated and grown for 14 days on soil before picked out and transferred individually to pots. Plants were grown under identical conditions in a controlled-environment growth room with (22°C, 8 h light, 16 h dark cycle). Arabidopsis thaliana *(Col-0) WT and mutant lines are shown. White scale bar represents 5 cm.*

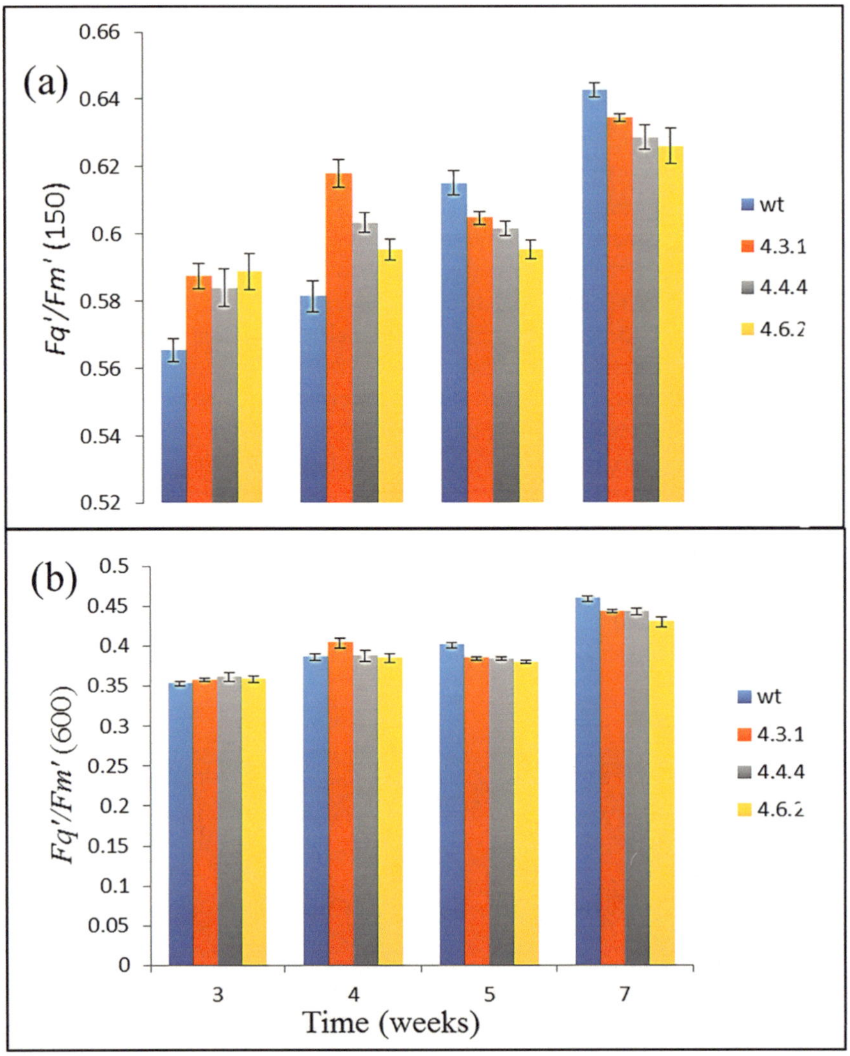

Figure 6.
Chlorophyll fluorescence imaging comparison of WT and mutant lines of construct 4-(pL2B-BAR-(pKST1)-AtRieske-tHSP to changes in light intensity. The maximum PSII operating efficiency (Fq'/Fm') values of the whole plant subjected to (a)150 µmol m -2 s-1 and (b) 600 µmol m -2 s-1 were measured. Plants were germinated and grown for 14 days on soil before picked out and transferred to individual pot on soil. These were grown for additional 5 weeks under identical conditions in a controlled environment growth room (22°C, 8 h light, 16 h dark cycle). Arabidopsis thaliana *(Col-0) WT and mutants individual lines of the construct 4-(pL2B-BAR-(pKST1)-AtRieske-tHSP measured. Data were obtained using 10–15 individual plants from three independent transgenic lines and are derived from weeks 3, 4, 5, and 7. Columns represent mean values, and standard errors are displayed, respectively. Significant differences between WT and lines ($P < 0.05$) at weeks 3 and 4 at 150 µmol m -2 s-. Each line was significantly different from wild-type WT. At higher light level of 600 µmol m -2 s-.1 no difference between the wild type and transgenic found in all the lines.*

showed a faster growth phenotype with larger rosettes in the transgenic lines than the wild type.

Growth was slower in the wild type in the early stages as seen above. However, as the plants advanced (week 7), the reductions in rosette area of the wild type seemed to catch up with the transgenic lines, which might again be implying that the expression of these two genes might be more active in early developmental stages.

7. Discussion

Transgenic studies have provided numerous evidence that manipulation of certain enzymes is potential route for the improvement of plant productivity [14, 28, 31, 36, 38, 41, 42].

In this work, we generated transgenic Arabidopsis (*Arabidopsis thaliana*) plants overexpressing the Rieske enzyme mutant lines with altered manipulation identified in *Arabidopsis thaliana*. The analysis carried out such as the DNA and herbicides screening plus the localization of the YFP have all revealed that these genes are present in the chloroplasts of the guard cells and had impact on the transgenic lines.

Phenotypes observed suggested that the transformed plants exhibited significant faster initial growth rates evidenced by larger leaf area and faster rosette increases which may suggest that Rieske might be of importance for enhanced plant growth. Studies have reported of increased yield in plant productivity by overexpressing these genes in the whole plants [28, 30, 36, 41]; however, it is interesting that we have observed similar enhancement of growth despite expression being limited to guard cells. The quantum efficiency of PSII photochemistry in the transgenic lines was also significantly higher in early development. These data imply that the photosynthetic efficiency of young plants may have a greater impact on plant development. These

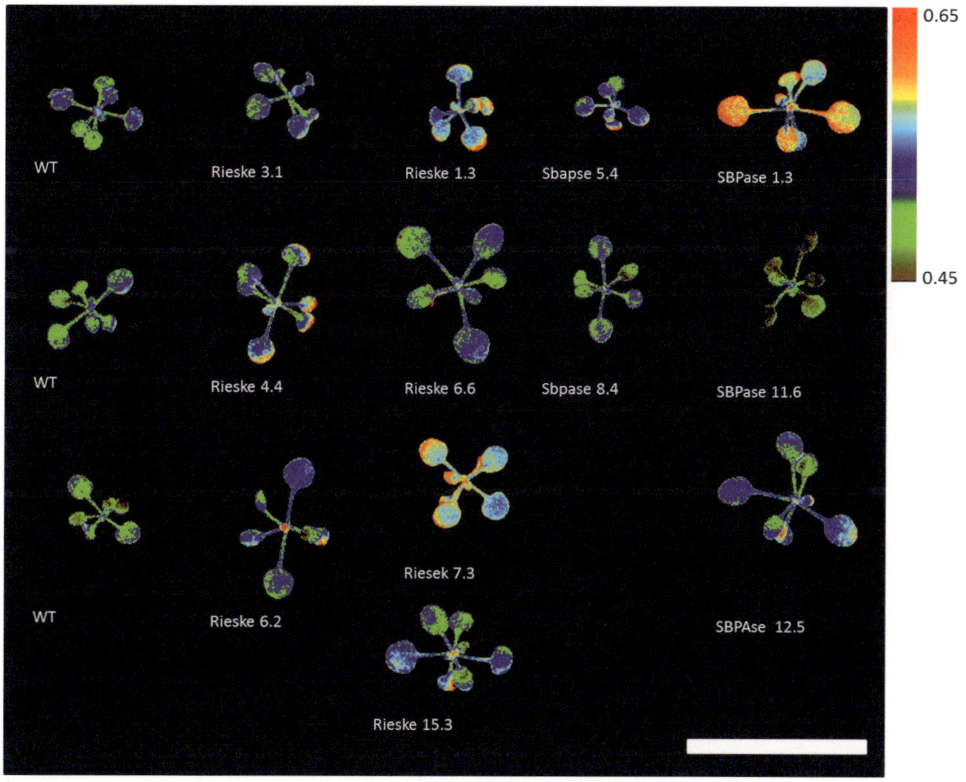

Figure 7.
Photosynthetic efficiency of PSll operating systems of images captured of WT and homozygous mutant Arabidopsis plants. 3–weeks old plants were germinated and grown for 14 days on soil before picked out and transferred to individual pots on soil. Plants were grown under identical conditions in a controlled-environment growth room with (22°C, 8 h light, 16 h dark cycle). Arabidopsis thaliana *(Col-0) WT and mutants are shown. Scale bars represent 5 cm and* Fq′/Fm′ *values represented by colors as indicated.*

findings are consistent with earlier studies, which reported that the stimulatory effects of increased levels of SBPase occurred earlier in development [28], which may also demonstrate the different limitations witnessed on photosynthesis between developing and fully expanded leaves.

It is important also to keep in mind that little changes in photosynthetic capacity counts and can have a great impact on plant development [28]. Therefore, these results suggest that altered expression (assumed due to the expression of the construct) Reiske in guard cells alone in plants seem to improve the overall plant photosynthetic efficiency and growth in these plants compared with the wild type suggesting that genes manipulation in the guard cells may be playing roles in plant development.

In conclusion, numerous studies have shown that photosynthetic enzymes in carbon metabolism have yielded increased photosynthetic rates in plant at the whole leaf level [34, 36, 41–45].

This research was based on elucidating the role of guard cells chloroplasts in stomatal regulation, and the role these cells could play in enhancing plant productivity. A comprehensive understanding of the signals or metabolites synchronizing stomatal conductance and carbon assimilation is, therefore, paramount toward successful manipulations of stomatal behavior for enhancing water use efficiency and sustainable inputs in agriculture.

The molecular approach efforts to comprehend the involvement of guard cell photosynthesis in stomatal function required manipulating photosynthesis specifically in guard cells, which demonstrated the potential of transgenic plants with altered guard cell metabolism. We particularly have demonstrated specificity of the KST promotor and shown that expression was only in the guard cells. This has also shown the potential of this promotor for manipulating guard cell-specific metabolism.

8. Additional information

Parts of this chapter were previously published in a doctoral thesis by the same author titled "The role of guard cell chloroplasts in stomatal function and coordinating stomatal and mesophyll responses" at the University of Essex, UK 2018.

Author details

Batta Kucheli
Adamawa State University, Mubi, Adamawa, Nigeria

*Address all correspondence to: battakucheli@gmail.com

IntechOpen

© 2024 The Author(s). Licensee IntechOpen. This chapter is distributed under the terms of the Creative Commons Attribution License (http://creativecommons.org/licenses/by/3.0), which permits unrestricted use, distribution, and reproduction in any medium, provided the original work is properly cited.

References

[1] Bergmann DC, Sack FD. Stomatal development. Annual Review in Plant Biology. 2007;**58**:163-181

[2] Buckley TN. The control of stomata by water balance. New Phytologist. 2005;**168**:275-292

[3] Hetherington AM, Woodward FI. The role of stomata in sensing and driving environmental change. Nature. 2003;**424**:901-908

[4] Meidner H, Willmer C. Circadian rhythm of stomatal movements in epidermal strips. Journal of Experimental Botany. 1993;**44**:1649-1652

[5] Lawson T, Blatt MR. Stomatal size, speed, and responsiveness impact on photosynthesis and water use efficiency. Plant Physiology. 2014;**164**:1556-1570

[6] Smith AM, Stitt M. Coordination of carbon supply and plant growth. Plant and Cell Environment. 2007;**30**:1126-1149

[7] Boyer JS. Plant productivity and environment. Science. 1982;**218**:443-448

[8] Gotow K, Taylor S, Zeiger E. Photosynthetic carbon fixation in guard cell protoplasts of *Vicia faba* L. evidence from radiolabel experiments. Plant Physiology. 1988;**86**:700-705

[9] Outlaw WH, Mayne BC, Zenger VE, Manchester J. Presence of both photosystems in guard cells of *Vicia faba* L implications for environmental signal processing. Plant Physiology. 1981;**67**:12-16

[10] Shimazaki K-I, Gotow K, Kondo N. Photosynthetic properties of guard cell protoplasts from *Vicia faba* L. Plant and Cell Physiology. 1982;**23**:871-879

[11] Zeiger E, Armond P, Melis A. Fluorescence properties of guard cell chloroplasts evidence for linear electron transport and light-harvesting pigments of photosystems I and II. Plant Physiology. 1981;**67**:17-20

[12] Zemel E, Gepstein S. Immunological evidence for the presence of ribulose bisphosphate carboxylase in guard cell chloroplasts. Plant Physiology. 1985;**78**:586-590

[13] Shimazaki K-I, Okayama S. Calvin Benson cycle enzymes in guard-cell protoplasts and their role in stomatal movement. Biochemie und Physiologie der Pflanzen. 1990;**186**:327-331

[14] Lawson T, Oxborough K, Morison JI, Baker NR. Responses of photosynthetic electron transport in stomatal guard cells and mesophyll cells in intact leaves to light, CO_2, and humidity. Plant Physiology. 2002;**128**:52-62

[15] Daloso DM, Williams TC, Antunes WC, Pinheiro DP, Müller C, Loureiro ME, et al. Guard cell-specific upregulation of sucrose synthase 3 reveals that the role of sucrose in stomatal function is primarily energetic. New Phytologist. 2015;**209**(4):1470-1483

[16] Schwartz A, Zeiger E. Metabolic energy for stomatal opening. Roles of photophosphorylation and oxidative phosphorylation. Planta. 1984;**161**:129-136

[17] Shimazaki K-I, Zeiger E. Cyclic and noncyclic photophosphorylation in isolated guard cell chloroplasts from *Vicia faba* L. Plant Physiology. 1985;**78**:211-214

[18] Madhavan S, Smith BN. Localization of ribulose bisphosphate carboxylase

in the guard cells by an indirect, immunofluorescence technique. Plant Physiology. 1982;**69**:273-277

[19] Frechilla S, Talbott LD, Zeiger E. The blue light-specific response of *Vicia faba* stomata acclimates to growth environment. Plant Cell Physiology. 2004;**45**:1709-1714

[20] Zeiger E, Zhu J. Role of zeaxanthin in blue light photoreception and the modulation of light-CO_2 interactions in guard cells. Journal of Experimental Botany. 1998;**49**:433-442

[21] Baker NR, Oxborough K, Lawson T, Morison JI. High resolution imaging of photosynthetic activities of tissues, cells and chloroplasts in leaves. Journal of Experimental Botany. 2001;**52**:615-621

[22] Lurie S. Photochemical properties of guard cell chloroplasts. Plant Science Letters. 1977;**10**:219-223

[23] Lawson T, Oxborough K, Morison JI, Baker NR. The responses of guard and mesophyll cell photosynthesis to CO_2, O_2, light, and water stress in a range of species are similar. Journal of Experimental Botany. 2003;**54**:1743-1752

[24] Melis A, Zeiger E. Chlorophyll a fluorescence transients in mesophyll and guard cells: Modulation of guard cell photophosphorylation by CO_2. Plant Physiology. 1982;**69**:642-647

[25] Kucheli B. Construction of sedoheptulose-1, 7-Bisphosphatase (Sbpase) for manipulation in guard cells of *Arabidopsis thaliana* L. International Journal of Plant & Soil Science. 2021;**33**:191-199

[26] Müller-Röber B, Ellenberg J, Provart N, Willmitzer L, Busch H, Becker D, et al. Cloning and electrophysiological analysis of KST1, an inward rectifying K+ channel expressed in potato guard cells. The EMBO Journal. 1995;**14**:2409

[27] Wang Y, Noguchi K, Ono N, Inoue S-I, Terashima I, Kinoshita T. Overexpression of plasma membrane H+-ATPase in guard cells promotes light-induced stomatal opening and enhances plant growth. Proceedings of the National Academy of Sciences. 2014;**111**:533-538

[28] Lefebvre S, Lawson T, Zakhleniuk OV, Lloyd JC, Raines CA, Fryer M. Increased sedoheptulose-1,7-bisphosphatase activity in transgenic tobacco plants stimulates photosynthesis and growth from an early stage in development. Plant Physiology. 2005;**138**:451-460

[29] Raines CA. The Calvin cycle revisited. Photosynthesis Research. 2003;**75**:1-10

[30] Simkin AJ, Mcausland L, Lawson T, Raines CA. Overexpression of the RieskeFeS protein increases electron transport rates and biomass yield. Plant Physiology. 2017;**175**:134-145

[31] Lawson T, Lefebvre S, Baker NR, Morison JI, Raines CA. Reductions in mesophyll and guard cell photosynthesis impact on the control of stomatal responses to light and CO_2. Journal of Experimental Botany. 2008;**59**:3609-3619

[32] Stroebel D, Choquet Y, Popot JL, Picot D. An atypical haem in the cytochrome b(6)f complex. Nature. 2003;**426**:413-418

[33] Yamashita E, Zhang H, Cramer WA. Structure of the cytochrome b(6)f complex: Quinone analogue inhibitors as ligands of Heme c(n). Journal of Molecular Biology. 2007;**370**:39-52

[34] Long SP, Zhu XG, Naidu SL, Ort DR. Can improvement in photosynthesis increase crop yields? Plant, Cell & Environment. 2006;**29**:315-330

[35] Raines CA. Transgenic approaches to manipulate the environmental responses of the C3 carbon fixation cycle. Plant, Cell & Environment. 2006;**29**:331-339

[36] Raines CA. Increasing photosynthetic carbon assimilation in C3 plants to improve crop yield: Current and future strategies. Plant Physiology. 2011;**155**:36

[37] Zhu X-G, De Sturler E, Long SP. Optimizing the distribution of resources between enzymes of carbon metabolism can dramatically increase photosynthetic rate: A numerical simulation using an evolutionary algorithm. Plant Physiology. 2007;**145**:513-526

[38] Simkin AJ, Mcausland L, Headland LR, Lawson T, Raines CA. Multigene manipulation of photosynthetic carbon assimilation increases CO_2 fixation and biomass yield in tobacco. Journal of Experimental Botany. 2015;**66**:4075-4090

[39] Engler C, Kandzia R, Marillonnet S. A one pot, one step, precision cloning method with high throughput capability. PLoS One. 2008;**3**:e3647

[40] Weber E, Engler C, Gruetzner R, Werner S, Marillonnet S. A modular cloning system for standardized assembly of multigene constructs. PLoS One. 2011;**6**:e16765

[41] Simkin AJ, Lopez-Calcagno PE, Davey PA, Headland LR, Lawson T, Timm S, et al. Simultaneous stimulation of sedoheptulose 1,7-bisphosphatase, fructose 1,6-bisphophate aldolase and the photorespiratory glycine decarboxylase-H protein increases CO_2 assimilation, vegetative biomass and seed yield in *Arabidopsis*. Plant Biotechnology Journal. 2017;**15**:805-816

[42] Von Caemmerer S, Furbank RT. Strategies for improving C4 photosynthesis. Current Opinion in Plant Biollogy. 2016;**31**:125-134

[43] Driever SM, Simkin AJ, Alotaibi S, Fisk SJ, Madgwick PJ, Sparks CA, et al. Increased SBPase activity improves photosynthesis and grain yield in wheat grown in greenhouse conditions. Philosophical Transactions of the Royal Society B-Biological Sciences. 2017;**372**:372-382

[44] Miyagawa Y, Tamoi M, Shigeoka S. Overexpression of a cyanobacterial fructose-1,6−/sedoheptulose-1,7-bisphosphatase in tobacco enhances photosynthesis and growth. Nature Biotechnology. 2001;**19**:965-969

[45] Uematsu K, Suzuki N, Iwamae T, Inui M, Yukawa H. Increased fructose 1,6-bisphosphate aldolase in plastids enhances growth and photosynthesis of tobacco plants. Journal of Experimental Botany. 2012;**63**:3001-3009

Chapter 7

Extraction of Renewable Natural Pigments in Indonesian Cultures for Coloring Batik Fabrics

Saefudin and Efrida Basri

Abstract

Nowadays, trends of Indonesian batik have increased to replace synthetic pigments with renewable natural pigments (chlorophyll, tannin, and others) due to strong consumer demands for more natural products. Production of renewable natural pigments from plant biomass (e.g., wood wastes) has been developed and produced in liquid, paste, and dry forms. Natural pigment production not only increases their selling-price capacity but should also exhibit biological activities that are environmentally friendly and beneficial for health. Abundant sources of chlorophyll/tannin in wood wastes (including mangrove origins) from forestry industries could potentially substitute for synthetic pigments, which sound environmentally harmful. Wood waste utilization could encourage and create economic opportunities in forest villages, centers of mangrove wood industries, and small/medium-scale batik/weaving enterprises. Natural colors of chlorophyll/tannin seem preferred by coastal and inland motif batik crafters. Accordingly, exploring various natural sources of tannin is essential for fabric/batik coloring, safe for food/beverages, and not endangering health.

Keywords: utilization, natural pigments, batik fabrics, environmentally friendly, wood wastes

1. Introduction

Renewable natural pigments obtained from living creatures/biomass (e.g., plants), such as chlorophyll, tannin, and others, have attracted remarkable attention from the related industry and trade world for coloring purposes (especially textile/batik fabrics, food/beverage packaging, paper, etc). The demand for chlorophyll/tannin pigments tends to increase continuously, as they are renewable, beneficial for human health; seem more environmentally friendly; and also due to the people's growing awareness of allegedly the possible dangerous compounds in synthetic pigments. Chlorophyll is a complex organic compound containing C, N, H, O, and Mg elements, present in almost any plants (particularly leaves) that functions as a biocatalyst in photosynthesis, whereby CO_2 and water (H_2O) are combined together to produce very simple food (sugar/hexoses). Further, the chlorophyll could exhibit bluish-green colors. Such color appearances could be due to the presence of double bonds (e.g., –C=C– in conjugation or not, –C=O) and unsaturated aromatic rings with N elements [1].

Meanwhile, tannin belongs to polyphenols (with C, H, and O elements), regarded as one of the indirect products of plant photosynthesis, which is formed afterward through assimilation and other complex metabolism processes. Further, tannin is commonly found in particular plant tissues, for example, twigs, leaves, barks, woods, flowers, fruits, and roots. Further, the physiological functions of the tannin presence in the plants are still unknown, possibly protecting the plant tissues against microorganism attacks [2]. The tannin could reveal brownish-yellow to reddish-brown colors, which could be attributed to phenolic, benzene, C=O, and conjugated –C=C– bond groups. Further, such presences in tannin as well as in chlorophyll enable both pigments to absorb light at a particular visible (colored) wavelength range (4000–8000 A), thereby exhibiting various and specific colors [1, 3]. Accordingly, chlorophyll and tannin could play essential roles in the coloring culture of weaving and crafting textile materials (especially the batik fabrics) and others (paper, food/beverage packaging, etc.) in Indonesia.

Batik makers, weavers, and eco-printers can look for chlorophyll and tannins around their homes. Color variations are different for each type and part of the plant, but the performance can be adjusted according to the customer's order. The presence of natural pigments is often together with other pigments such as flavonoids, anthocyanins, and carotenoids [4].

Chemically chlorophyll contains polar groups (e.g., CHO, Mg^{+2} ions, and COO); however, its large nonpolar portions are insoluble in water. Further, plants commonly have two types of chlorophyll (i.e., a and b). Chlorophyll b is more soluble (in water and other polar solvents) than a because the former contains aldehyde (CHO) groups, while the latter is not so. Meanwhile, the tannin is only partially water-soluble, owing to the presence of hydrolyzable (more hydrophilic) structures and condensed/nonhydrolyzable (less hydrophilic) structures in particular proportions. However, chlorophyll and tannin are soluble in alcohols, ethers, and hot water (moreover, with the addition of detergent or alkali).

Regarding other plant pigments (e.g., flavonoid, anthocyanin, and carotenoid), they mostly could be, among others, polyphenols (with highly condensed structures), and the compounds with the presence of double bonds (–C=C– and –C=O) and phenolic groups, as well as macromolecules (with sizable nonpolar portions, despite containing polar groups). Consequently, most of them are insoluble or only partially soluble either in water. However, as occurred to chlorophyll/tannin, those other pigments could be expectedly solubilized in hot water. This is because the hot water could induce more kinetic molecular movement and hence cause more intensive melting of pigments, thereby enhancing their water solubility.

The solubility properties and other characteristics of chlorophyll/tannin as well as other pigments are essential for coloring purposes (e.g., textile/batik fabrics, paper products, and food/beverage packaging). This is because the coloring works, beginning from the pigment extraction (from the plants), the coloring processes, until the finalization, usually employ aqueous liquid media. Further, the presence of double bonds, carbonyl (–C=O), and phenolic groups in those other pigments, as also the case in chlorophyll/tannin, renders them to also absorb visible light at a particular wavelength and hence to appear colored [1, 4]. Further, similar to tannin, those other pigments (e.g., flavonoids, saponine, and anthocyanin) could possibly be formed as indirect products of photosynthesis through further assimilation/other complex metabolisms. The function of those pigments for the plant themselves might be, among others enhancing the plant-tissue durability, releasing specific smells (for the arrival of wanted animals/insects), and attracting the insects to come (especially those in the flowers with their attractive colors) to enhance the plant/flower pollinations. Pollination occurs mostly in the flowers, which refers to the transfer of pollen grains to the surface of the stigma, aiming mainly to develop future seeds. Those seeds would further develop into young plants and ultimately mature plants. As the plants grow mature/older, the reserve food (resulting from photosynthesis) stored mostly in the plant parenchyma tissues might exceed the needs for plant assimilation and other regular metabolism activities. This situation renders the excess food in the parenchymas to be transformed (under no or little oxygen) into the so-called plant extractives, including also the synthesized natural pigments [2].

Communities in forest villages can produce chlorophyll and tannins from plants such as betel nuts (*Areca catechu*), jackfruits (*Artocarpus heterophylla*), bananas (*Musa paradiciaca*), mangroves (*Rhizophora mucronata*), pine (*Pinus mercusii*), and *Uncaria gambir*. Those in the highlands can get it from the tea plant (*Camelia sinensis*), while those in the lowlands can get it from *Rhizophora mucronata, Xylocarpus granatum, Avicenia marina, and Terminalia catappa*. Especially in the coastal area are the species of *Rhizopora apiculata, R. macrophylla, Pterospermum ferroginum, Ceriops candolleana*, etc.

The source of tannin and chlorophyll comes from tree waste which was produced as much as logging forests, amounting to around 928,934.81 m^3/year [5]. Another source of log production from plantation forests reaches 40,945,378.90 m^3 per year (Ministry of Environment and Forestry of the Republic of Indonesia/Keputusan Menteri Lingkungan Hidup dan Kehutanan, 2018). The source of tannin, chlorophyll, and other pigments mainly comes from the bark of the mangium wood, which can reach 216.49 tons per day or around 9% of logs by weight [6].

Further, wood wastes in the forms of wood slabs, sawdust, barks, etc., are supposedly rich in various natural pigments, such as tannin, flavonoid, anthocyanin, chlorophyll, and carotenoid. So far, those alleged pigment-rich wood wastes are still not utilized optimally, instead just abandoned or discarded. Consequently, the waste utilization into renewable natural pigments could be a better solution/recommendation and hence deserve consideration.

In obtaining/producing renewable natural pigments, after their extraction from plant biomass portions (e.g., barks, woods, roots, leaves, and flowers/fruits) or wood wastes, they could be produced in three forms/shapes (i.e., dry-solid extracts, concentrated pastes, or liquid shapes). Whatever their shapes, attempts should be thoroughly performed, thereby enabling the pigment to meet the standards of quality, security, and uses or benefit principles. Further, the pigment's raw materials should be consistently available and easily assessed. The pigment products should bear their specification, especially the color standard, color consistency, information about the ingredients as well as other matters added in the pigment formulation, and essentially their shapes. Accordingly, pigment testing deserves to be carried out.

Regarding the shapes/forms of pigment products, however, for the sake of efficiency in long-termed use, especially the trade, it would be better to process/produce the chlorophyll/tannin as dry-solid pigments. Dry natural pigments could last longer or be more durable during storage, rapidly packaged, and easy in case of coloring work/activities (just dissolve them in the water). The essential coloring activities (chiefly for the fabrics/batik) with the pigments, for example, chlorophyll/tannin, consist mainly of mordanting, coloring, and fixating. Mordanting aims to eliminate fat, oil, and other foreign matters, which could adhere or stick to the fabric fibers, thereby otherwise interfering with the fiber-pigment bonds. Also, such elimination could greatly facilitate pigment infiltration/penetration into the pores, voids, or other microscopic fiber structures (micropores/microvoids) during the fabric coloring. Mordanting is usually performed before the fabric coloring, using typol, tawas/alum, tannic acid, and acetic acid/vinegar (all in aqueous solution). Fabric coloring is carried out by immersing the mordanted fabrics in an aqueous coloring-pigment solution (e.g., chlorophyll/tannin). Afterward, the fixation could be conducted, whereby the pigment-colored fabrics are submerged (immersed) in the fixative solution. The fixation intends to strengthen the color fixing at the pigmented fabrics [7]. Further, to examine the pigment performance (e.g., chlorophyll/tannin) at the pigment-colored fabrics, then their color-leaching/fading resistance should be tested against washing, rubbing, and exposure to sunlight, which refers to the recognized standards (e.g., ISO/international standard organization and SNI/Indonesia's national standards).

Accordingly, research into diverse natural sources of pigments (e.g., tannin and chlorophyll) that are safe for food/drink as well as the human body is crucial for fabric and batik coloring. Relevantly, this chapter presents, elaborates, and assesses the results of exploration, extraction, and application of natural pigments (e.g., chlorophyll/tannin and others) extracted from various plant species/portions/wood wastes and motivation of batik, weaving, and eco-print crafters.

2. Current main situations and elaborated approaches

Natural renewable pigments now become one of the prominent complements in substituting for synthetic pigments for coloring purposes, such that the coloring results become the community's partial lifestyle. Natural pigments from wherever they come are preferred a lot by consumers (users) because of their superiority, which is inherently exclusive; renewable; environmentally friendly; and appear very typical, distinct, and ethnic, thereby affording high selling values. Natural pigments could become potential superior products regionally and globally. Consequently to achieve such, it is necessary to conduct research in order to get even better results from pigment extracts as well as pigment-colored products (e.g., especially, fabrics/batik, paper, and food/beverage).

The research which have been performed partially revealed that many ethnic groups throughout the world have utilized particular plant species/portions (and vegetation/wood wastes) in natural renewable pigments, including in Indonesia. For such, Indonesia's ethnic groups have utilized specific plant biomass, among others: ketapang roots (*Terminalia catappa* L), teak leaves (*Tectona grandis* Lf.), barks of *Rhizophora* sp., annato seeds (*Bixa orellana* L.), tea leaves (*Camelia sinensis* L), and betels nuts (*Areca catechu* L), as they feel concerned with securing the environments. Meanwhile, several countries, such as India, Singapore, Malaysia, and China, have conducted intensive research and begun performing the trade transaction dealing with renewable natural pigments because they mostly get attracted to environmentally friendly and renewable products (e.g., plant pigments). Relevant to such favorable achievements, this still needs to be enhanced toward substituting natural pigments for synthetic pigments, as elaborated in the following related essential topics, respectively.

3. Production of renewable natural pigment

In the production aspects, it is essentially related to how renewable natural pigments are obtained. Renewable natural pigments (e.g., chlorophyll, tannin, flavonoid, saponine, carotenoid, and anthocyanin) could be obtained commonly through the extraction process from almost any parts of the plant tissues/biomass allegedly rich in such pigments, for example, tree barks, roots, leaves, fruits, seeds, exudates/saps, and fruit skins [8, 9]. The terms extraction and production are often used interchangeably. In the extraction/production of natural pigments (e.g., chlorophyll/tannin), currently, the particular plant portions considered worth for such are the wood wastes generated abundantly from the tree cutting/felling in the forests and from the wood sawing in the sawmilling industries. So far, those wastes are still underutilized, instead as described before, just abandoned in the forests, merely left accumulated on wood-industry sites, or just burnt uncontrollably, thereby potentially polluting the environment and hence arousing the idea for waste utilization for the production of renewable natural pigments.

Further, those would-be natural pigments (e.g., tannin, flavonoids, saponine, anthocyanin, chlorophyll, and carotenoid), besides already naturally present in the plant biomass or wood wastes, could also be accidentally formed during their heating or storage. This is because during such, chemical and biological changes could occur, such as degradation, depolymerization, respiration, fermentation, and other physiological processes, thereby chemically converting the more complex and higher molecular weight (MW) plant components to simpler and lower MW pigments [10]. Accordingly, those phenomena could also affect or contribute to the pigment productions.

The production of chlorophyll/tannin and other pigments should be preceded by the exploration of the allegedly pigment-rich plants (or wood wastes) which grow or exist in the surrounding community. The simplest way of performing the exploration is through the identification process using specific chemicals. For example, the extracts from the fruits, leaves, and stem barks of mangrove (*S. alba*) trees, after identification using phytochemical tests with $FeCl_3$, turned their colors to blackish green or strong blue, which indicated the tannin presence. Still relevant, the barks of *S. alba* stems, after being dried, maintained their colors similar to those before being dried but then exhibited dry and hard textures [11, 12]. Unfortunately, still, not many of the extracted pigments (e.g., tannin) are ready for use (usable), continuous, and standardized, and nor is their availability in the markets.

To cope with such pigment drawbacks (e.g., its ready use, incontinuity, standardization, and market availability), among the ways, it is necessary to enhance the pigment yields that result from the extraction. This is because the pigment yields are affected by the species origins of the plant biomass (e.g., wood and wood wastes) and the kind of cooking/extracting chemicals. For example, this occurred, when woods of various mangrove species were extracted in hot water (near 100°C), other polar solvents (alcohol/acetone), or aqueous alkali solution (0.5–1.0% Na_2CO_3). It turned out that the extraction using 0.5–1.0% Na_2CO_3 (alkali) afforded higher chlorophyll and tannin yield, viscosity, and solid content than using water. As such, the highest solid content (7.866%) in the tannin was obtained from the woods of *A. officinalis*, using a 1.0% Na_2CO_3 solution. Meanwhile, correspondingly the highest tannin yield (31.003%) was found at *R. apiculata* woods [13]. These results indicated again chlorophyll and tannin pigments contained less water-soluble portions, more sizable less-polar structures, and greater molecular weight compounds, which could only be effectively solubilized in hot water or aqueous alkali solution. The data/information about the pigment yields (e.g., tannin) is essential because it is closely related to the pigment potency.

The potency of natural pigments (e.g., chlorophyll/tannin) in particular plants growing at particular area could be predicted from the density of plant stands there (e.g., number of stands per ha; or average biomass weight of the overall stands per ha). As such, as an example, research has been conducted on the tannin pigment in the plant stands that grew on rehabilitated lands at Rembang (Central Java). First of all, the tannin content/yield in the plant biomass was determined (w/w) through a laboratory experiment. Further, by measuring the stand density (i.e., number of stands per ha or stand-biomass weight per ha), then the tannin potency could be approximated/predicted, that is, roughly 105.93 kg/ha/year. Still related, after the third year of stand growth, the tannin potency reached 146.36 kg/ha/year [14]. Consequently, knowing the potency of natural pigments (e.g., tannin) is essential because, further, the tannin contains specific compounds that could potentially exhibit exclusive tastes/smells.

Among such exclusive tastes/smells is that the tannin could render its compounds to taste bitter and astringent, which could be toxic, thereby besides protecting the plant tissues against organisms (as presumed before), also decreasing the plant digestibility and hence deterring the scavenging herbivores and other pests to consume them. Such bitter/astringent tastes are often found in young fruits. Changes in tannin compounds could exert a significant role in fruit ripening. Accordingly, it is strongly alleged physiologically, perhaps that tannin serves as plant-growth regulator. Further, tannin content in the plant biomass (e.g., slabs, twigs, and woods), inadvertently dissolved and leached by the rainwater (with various kinds of humus), renders the colors of the flooded/stagnant water in the swamps or peat marshes to appear blackish brown like those of tea water, popularly known as black water. The tannin also causes bitter/astringent tastes and imparts dark brown colors to the tea water.

Relevantly, imparting the colors by natural pigments (e.g., tannin/chlorophyll and others) at the fabrics (e.g., batik) should proceed in stages. This is because most of the pigments are, as described before, insoluble or only partially so in the water. Therefore with stages, it could render the pigments more water-soluble. Moreover, in the coloring works, as has been described, it employs mostly aqueous liquid media. The first stage in the fabric coloring (after the fabric cleaning and mordanting) starts with immersing the cleaned/mordanted fabrics in the aqueous pigment solution, followed by the fixation stage. In this stage, the pigmented fabrics are fixated with specific inorganic salts such as kapur/lime ($CaCO_3$), tawas/alum ($Al_2[SO_4]_3$), and tunjung/ferrous sulfate ($FeSO_4$). Fixing agents direct chlorophyll and tannins into light or dark colors and are more resistant to washing, sun exposure, and ironing. The common procedures in coloring the fabrics are as follows: 50 grams of any chemical fixatives (lime, alum, or tunjung) are dissolved in 1 liter of water. The resulting fixative solution is then allowed to settle/precipitate for 24 hours until two layers are formed (upper and lower). The upper/supernatant layer is further used as the fixative solution. The fixation is performed by immersing the pigmented fabrics in the fixative solution for approximately 15 minutes. Afterward, the fabrics are removed, allowed to dry, and ready for color detection. As such, the dried fabrics are immersed again in the specific color-detecting aqueous liquid. The color detection/identification test procedures refer to the RHS (royal horticultural society) color charts. Afterward, the color-identified fabrics are properly documented by inscribing specific labels, such as kinds of pigments, kinds of fixatives, appearances of the imparted colors, 2022.

As of this occasion, despite strenuous attempts in imparting various pigment colors/motifs at the materials (e.g., fabrics) with the fixative aids, unfortunately, those available varying sources of renewable natural pigments which are abundant in Indonesia and more environmentally friendly have not yet provided maximal solution, especially as a substitute for synthetic pigments. Further, in spite of strong suggestions and appeals for Indonesia's batik crafters/weavers to use natural pigments, based on such positive allegations, pitifully, in fact, the majority of domestic batik crafters/weavers still use synthetic pigments. Accordingly, systematic and thorough attempts deserve carried out to reduce the dependence on imported synthetic pigments, among others, by establishing renewable natural pigment industries that utilize chlorophyll (green), anthocyanin (red-violet), xanthophyll (yellow), tannin (brown, reddish brown), and carotene (orange), which are abundantly available or existed in the plants.

Further, in order that Indonesia's abundant natural pigments could be obtained and produced into usable or ready-for-use pigments in feasible operation (technically as well as economically), then it is necessary to explore the genetic and biochemical aspects of the sources of natural pigments (especially the plants), which can bring essential economy values. Accordingly, a breakthrough is urgently performed with respect to the effectiveness, efficiency, and rules to regulate the utilization of those natural biopigment sources. In addition, the involvement of the batik-crafting community and traditional fabric weavers should get a priority. The aims are that the pigment-industry continuation and environment sustainability could be realized.

4. Variations in colors and motifs at batik fabrics

Batik fabrics are one of the pigment-colored and motif-decorated products. As such, the traditions performed by the batik crafters and fabric weavers to impart the fabric colors and motifs using natural pigments are by cooking (boiling in the water) the plant biomass (e.g., barks, leaves, wood slabs, sawdust, and flowers), until the colors appear in the boiled water (as pigment solution). When the colors become thickened or viscous enough, the fabrics are immersed in the pigment solution to undergo the coloring, as systematically shown in **Figure 1**. These local coloring procedures are conducted a lot by the forest village's community, which further become a guidance/reference, adopted by consecutively the regional authority, Indonesia's state-owned forest enterprise (Perhutani) in Central and East Java; and wood enterprises in Kalimantan, Sumatra, Nusa Tenggara, and Bali.

The colors of natural pigments (e.g., chlorophyll, tannin, and others) become essential to seek and arrange the motifs at the pigmented fabrics, which should be compatible with the desires of the batik/fabric crafters; and also with the tastes of the ordering users/consumers. If we want motifs other than chlorophyll and tanning

Figure 1.
Illustration of coloring the fabrics, using the pigments extracted from consecutively leaves dan barks of avocado (Persea americana Mill), jambu seeds (Psidium guajava L.), mahoni (Swietenia mahagoni; Jacq), manggous (Mangifera indica L.), and coffee (Coffea arabica L), (using the fixatives: Alum, lime, and ferrous sulphate).

pigments, we can use waste from plant species or other sources of wood waste. Plant biomass and wood wastes produce a variety of colors. The leaves were dominated by chlorophyll can create green. The presence of flavonoids can create yellow. If tannins dominate the leaves and bark, they give brown, brownish-yellow, or reddish-brown [15, 16].

The varying colors imparted by those several kinds of natural pigments should be confirmed more convincingly with regard to the pigment source origins and their positive presence. As such, source origins that cover tree roots, barks, stems, and leaves from mangrove (*Avicennia* sp.) plants are prevalently very rich in chlorophyll, tannin, phenolic, and flavonoid compounds, where their presence is indicated, after the detection tests, by a strongly positive category [++] to a very strongly positive category [+++] [17, 18]. This implies that if a part of those mangrove plants is dissolved in hot water, then the resulting colors could appear very strong, that is, blackish brown, resembling the tea-water colors. The presence of particular pigments (e.g., chlorophyll, tannin, and flavonoid) with high content in mangrove plant parts allows the extracts to be used as a natural pigment for fabric coloring.

Besides mangroves, other plant species (e.g., old teak trees) could also exhibit specific colors, which were due to the presence of extractives, predominantly 2-methyl anthraquinone (as polyphenols), where its content was quite great (reaching 13.54%). In teak wood, the quinones together with wood lignin allegedly enhance wood durability and also imparts attractive appearance/color to wood surface as well as seemingly renders easier the wood working (e.g., sanding, planing, and polishing) into favorable teakwood products [19]. However, the teak extractives contained tannin only a little [20, 21]. Further, the tannin content in the extracts from *Acacia mangium* barks could reach 15–20% [22]. The batik fabrics which were colored with *Acacia amngium*'s tannin, were resistant against the high-intensity sunlight and also against the pH changes [15].

Likewise, coloring the cotton fabrics with the extracts from teak (*Tectona grandis*) leaves, previously mordanted with alum and acetic acid (vinegar) separately, brought the fabrics with different color motifs, expressed in brightness values, that is, 55.56% (less bright) and 77.22% (brighter), respectively [23]. On other occasions, chlorophyll and tannin pigments were extracted from the barks of *Acacia mangium*, three mangrove species (*Terminalia catappa*, *Rhizophora apiculate*, and *R. mucronata*), and *Tectona grandis*, and then used concurrently for fabric coloring, followed with the fixation (using alum, lime, and ferrous sulfate, separately). It turned out that these procedures that incorporated chlorophyll and tannin could be used either as basic color motifs or as supplementing the colors of the resulting batik fabrics [24–26]

In imparting the fabric colors, it turned out that tannin is the most dominantly used in batik fabrics. Also, tannin is used in the coloring aspects and the color variations, where the two latter cases are the most important in designing and developing the batik colors and motifs, especially the inland motifs [27]. On other occasions, research results revealed that different colors in the colored fabrics could occur using particular pigments due to the use of different fixatives. This case was commensurate with the previous statement that the appearance of different colors in the pigmented fabrics could be due to different fixatives. The fixative use actually aimed to strengthen the bonds of fabric-fibers-to-pigments.

Further, research results in using a fixative solution, after the batik-fabric coloring, revealed that the fabric colors would not easily fade or leach out and become more resistant to rubbing/ironing [25]. The fixatives used in this research consisted of tawas/alum, kapur/lime, and tunjung/ferrous sulfate. Those three fixatives were used because they had been utilized a lot; moreover, the fixative prices were affordable, inexpensive, and easily obtained in the markets. Meanwhile, the natural pigments used for such were the chlorophyll and tannin extracts from *Acacia mangium* and two mangrove species (*Terminalia catappa* and *Rhizophora apiculate*).

Relevantly, tests have been conducted on 73 plant species, which supposedly could serve as natural pigment sources. It turned out 26 species were identified containing natural pigments (allegedly tannins) that imparted yellowish brown, reddish brown, or strong brown colors for basic colors of batik fabrics [1, 13, 14, 28]. Those tannins originated from the particular parts of those 26 species, such as leaves, barks, flowers, and twigs (**Table 1**).

The natural pigments (i.e., tannin) extracted from the root parts of those particular plant species (**Table 1**) for fabric coloring, then treated with three different fixatives (tunjung, alum, and lime) brought the fabrics with three different colors. As such, after being fixated on tunjung, the fabric colors appeared dark, in average black to gray colors. Meanwhile, correspondingly after being fixated with alum, the fabric colors became brighter than the original pigmented-fabric colors (without fixative treatment). Likewise after lime fixation, the pigmented-fabrics generated blackish colors [27].

The dark/blackish colors (**Table 1**) appearing at the pigmented fabrics (after tunjung fixation)) indicated that a reaction occurred between the fixative and tannin pigments (plant extractives) that produced complex salts. The complex

No	Species of the plant	Parts of the plants	Colors imparted/ appeared on the fabrics
1	*Acacia mangium*	Tree barks, leave	Strong brown, green
2	*Areca catechu*	Fruits, leave	Brown, reddish-brown, green
3	*Artocarpus heterophylla*	Tree barks, leave	Yellowish-brown, green
4	*Bixa orelana*	Tree barks, leave	Blackish-brown, green
5	*Caesalpinia sappan*	Tree barks, leave	Brown, Yellowish-red, green
6	*Camelia chinensis*	Leaves, tree barks	Green, Reddish-brown
7	*Ceriops candolleana*	Tree barks, leave	Yellowish-brown, green
8	*Cudraina javanensis*	Tree barks, leave	Reddish-brown, green
9	*Garcinia mangostana*	Fruit skin, leave	Brown, reddish-brown, green
10	*Morinda citrifolia*	Barks, leave	Brown, reddish-brown, green
11	*Pterocarpus indicus*	Tree barks, leave	Red, yellowish-brown, green
12	*Koordersiodendron pinnatum*	Leaves, twigs	Strong red, brown, green
13	*Musa paradiciaca*	Flowers, fruits	Brown, blackish-brown, green
14	*Peltophorum pterocarpum*	Tree barks, leave	Brownish-red, red-black, green
15	*Pinus mercusii*	Tree barks, leave	Reddish-brown, green

No	Species of the plant	Parts of the plants	Colors imparted/ appeared on the fabrics
16	*Psidium guajava*	Tree barks, leave	Greenish-brown, green
17	*Rhizopora apiculata*	Tree barks, leave	Brown, blackish-brown, green
18	*Rhizopora mucronate*	Tree barks, leave	Blackish-brown, brown
19	*Soneratia alba*	Tree barks, fruits	Brown, greenish-brown
20	*Syzygium conglomeratum*	Tree barks, leave	Brown
22	*Swietenia mahagoni*	Tree barks, fruit skins	Brown, red, yellowish-brown
23	*Terminalia catappa*	Tree barks, leaves	Brown, black, brownish-yellow
24	*Tectona grandis*	Leaves, tree barks, woods	Red, brown, blackish-brown
25	*Uncaria gambir*	Leaves, twigs	Strong red, brown
26	*Xylocarpus granatum*	Tree barks, fruits	Brownish red, brown

Sources: Jamal [29]; Basri et al. [30]; Saefudin et al. [1]; Susanto (1980); Prayitno et al. [28], Poedjirahajoe et al. [14]; Suhendry et al. [13]

Table 1.
Parts of the particular plant species, regarded as sources of tannin's natural pigments for coloring the batik fabrics and other woven fabrics.

salts, being large-sized compounds, could additionally provide a bridging (intermediary) between the pigment and the fabric fibers, thereby further creating strong fiber-pigment bonds (fixing) and, accordingly, imparting strong color adherence at the pigmented fabrics. Meanwhile, the different colors (not dark/blackish colors, but brighter colors) of the pigmented fabrics after the lime as well as alum fixations, supposedly did not produce complex salts, thereby rendering the fibers-pigment bonds not too strong. However, the lime as well as alum fixatives, could afford ionic bonding with either fabric fibers or pigments, thereby still imparting the fabric-pigment bonds, although not so strong as the bonds exerted by the complex salts [31].

Further, it appeared that the use of different fixatives (i.e., alum, lime, and tunjung/ferrous sulfate) separately at particular fabrics brought the pigmented fabrics with the colors toward the brightest, then less brightness and ultimately the darkest, respectively (**Table 1**). In another case, different kinds of fabrics consecutively with cellulose-based fibers (i.e., mori, cotton, and cotton yarns) and with protein-based fibers (wool), but colored with the same natural pigment and then treated with the same fixatives, brought the pigmented-fabrics with different colors. This difference occurred because mori/cotton fabrics and cotton yarns comprise mostly cellulose polymers, thereby rendering the fabric/yarn fibers to easily adsorb the pigments and fixatives. However, the wool (as the animal/sheep hair fibers) comprises mostly protein polymers, which could have different characteristics in interacting/adsorption with the pigments/fixatives. Accordingly, cellulose-based fabrics/yarns are widely used as pigment-colored media by batik crafters/weavers [32].

5. Opportunities and challenges

Several batik and eco-print artisans in Central Java and Yogyakarta have started cultivating, producing and selling natural chlorophyll and tannin pigments commercially. Available dry extracts include: Indigofera tinctoria, Camelia chinensis, Intsia bijuga. Other plants that are also rich in natural pigments (tannin and chlorophyll) are soga tingi (*Ceriops candolleana*), tegeran (*Cudraina javanensis*) and noni (*Morinda citrifolia*), jambal (*Pelthophorum ferruginum*), kesumba (*Bixa orelana*), and guava (*Psidium guajava*) (Susanto, 1980; [1]). Although those extracted natural pigments have been known a lot by fabric crafters/weavers; unfortunately, only a few have utilized them. The main obstacles for such few users are the limitation of pigment varieties that are ready for use (usable), lack of the crafter/weavers' knowledge about pigments' raw materials, and insufficiency in ready-for-use technology. Besides, the availability of natural pigments is affected much by the seasons, environments, and collecting workers, thereby causing a not-continuous pigment supply to the crafters/weavers. Despite their drawbacks, the natural pigments could prospectively afford high market potencies as Indonesia's superior commodity to enter into the global markets due to their attractiveness with unique, ethical, and exclusive characteristics. Those pigment drawbacks and prospects could become notable challenges for the batik crafters and related researchers.

The challenge for the researchers is that it is necessary to explore the natural pigment sources ready for use. The creativity of fabric crafters/weavers to seek fabric motifs that should conform to the user's tastes and desires could become their own problems. These classical problems also occur to the producers of batik and other woven fabrics scattered in many Indonesia's provinces. The batik crafters should obtain pigment-rich plants, which are fast growth, shortages, enormous available, and easily obtained in their home vicinities. To obtain the plant species rich in natural pigments with such favorable plant-growth characteristics necessitates thorough development for the present or future.

The development of natural pigments in the future should focus on research in the design motifs to anticipate local and global markets. Kinds of laboratory researches that use experiment methods should try hard to invent efficacious/unique formulae to develop such pigments in the crafter-community vicinity. Commonly, natural pigments are obtained/extracted by cooking the plant biomass/wood wastes in hot water (near 100°C). Further, to make the colors of the pigmented stuffs (e.g., fabrics) last long or as color protection against fading/leaching, it should use fixatives (alum, lime, tunjung, etc.). Plant biomass and wood wastes are allegedly rich in natural pigments (e.g., tannin/chlorophyll, anthocyanin, xanthophyll, and quinone). Evaluation results alleged that tannin's natural pigment was the most prospective for coloring purposes (e.g., fabrics).

Despite the reported success/prospects of using natural pigments (especially tannin), the data and information about the pigment potency (e.g., from forests/wood wastes) are still lacking or inadequate and have not yet been assessed. Furthermore, using tannin as a natural pigment from forests/wood wastes could become an added value for the tannin's plant sources and essential information for the community who love Indonesia's batik fabrics and other woven materials colored with the tannin.

The plant tannin, besides already being intended as fabric coloring, has long been widely used as a tanning agent to convert animal bones and skins into leathers (under high temperatures), which are stable, strong, and elastic/flexible. The tannin, as high molecular weight polyphenols, contains numerous OH and other polar groups (e.g., CO and COO) and is, therefore, able to impart multiple strong hydrogen bonds not only with fabric fibers but also with hydroxyl/other polar groups at the bone/skin proteins. Tannin presence in the plant leaves is concurrent with the chlorophyll presence, thereby allegedly causing the changes in leaf colors. The leaves of whatever plant species could also undergo the color changes sooner or later because the leaves perform a double function. As such, the leaf chlorophyll absorbs the red and blue colored light from the sunlight while concurrently conducting photosynthesis. Then, gradually reducing the capacity of chlorophyll to absorb sunlight, reducing the intensity of the green color of the leaves, until they become old, dry out and finally fall. Meanwhile, the carotene that imparts red colors to the leaves is more stable than the chlorophyll, thereby enabling the leaf's reddish-orange colors to stay longer, although the leaf's green colors disappear [33]. Other factors that could affect the pigment stability (e.g., their colors) of cations, oxygen, pH, sulfur dioxide (SO_2), proteins, and enzymes. All factors should then be considered.

Considering such, it is essential to know that the attractive varying colors imparted by plant's natural pigments (e.g., chlorophyll, anthocyanin, flavonoid, carotenoid, etc.) have attracted enthusiastic attention among many genetic experts. This is because such pigments might have meaningful relations with the various morphology aspects in plant species (but in the same genus), which are still close relatives. That information is essential for the taxonomist to determine and combine the plant's evolution lines into one group. Regarding flavonoid pigments, their form in the plants could be induced by the blue-colored light radiation (e.g., from the sunlight) to increase the plant resistance against ultraviolet radiation.

6. Conclusion and suggestions

Indonesia is endowed with abundant availability of renewable and environmentally friendly natural pigments, with respect to their huge potencies, numerous pigment kinds (e.g., tannin, chlorophyll, flavonoids, saponine, anthocyanin, and carotenoid), and their varying colors (e.g., green, red-violet, yellow, brown, reddish brown, and orange). Expectedly those could serve as substitutes for synthetic pigments, which are mostly environmentally unfriendly.

Those natural pigments could be obtained from almost any part of the plant biomass (e.g., tree barks, roots, leaves, fruits, seeds, exudates, and fruit skins) and wood wastes through commonly the extraction process, which is further beneficial for imparting colors and motifs to, for example, specially woven fabrics/batik, paper products, and food/beverage packaging, commonly performed by the synthetic pigments.

Unfortunately, the majority of Indonesia's users/consumers still use synthetic (inorganic) pigments, which are mostly imported. One way to overcome such is by establishing industries to produce natural pigments, which could last longer and be ready for use. Accordingly, the natural pigments (as the pigment-colored products, e.g., batik fabrics) should incorporate the fixation process (to strengthen the fabric fiber-pigment bonds), using, for example, alum, lime, and tunjung/ferrous sulfate fixatives); and undergo the quality tests (among others the color resistance against the leaching/fading, due to mainly the detergent washing, rubbing/ironing, and sunlight exposure).

The success in utilizing the plant biomass and wood wastes into pigments, which are not only confined to chlorophyll and tannin (currently the dominant natural pigment for especially batik coloring and tanning agents), but extended to others (flavonoids, carotenoid, anthocyanin, etc.), could be beneficial and safe for health, thereby expectedly encouraging and creating economic opportunities as well as welfare, for example, pigment industries in small/medium and large scale levels, users/consumers, and enthusiastic community.

Author details

Saefudin[1,2]* and Efrida Basri[1,2]

1 National Research and Innovation Agency, Research Center for Biomass and Bioproducts, Bogor, Republic of Indonesia

2 Cibinong Science Centre, Bogor (West Java), Indonesia

*Address all correspondence to: saefudinkahfi@gmail.com

IntechOpen

© 2023 The Author(s). Licensee IntechOpen. This chapter is distributed under the terms of the Creative Commons Attribution License (http://creativecommons.org/licenses/by/3.0), which permits unrestricted use, distribution, and reproduction in any medium, provided the original work is properly cited.

References

[1] Saefudin K, Basri E, Santosa A. Bahan Alam Pewarna Batik & Tenun. Bogor: Forda Press; 2018

[2] Starr C, Taggart R. Biology: The Unity and Diversity of Life. Fifth Edirion ed. Belmont, California, USA: Wadsworth Publishing Co; 1993

[3] Koch P. Utilization of Harwoods Growing on Southern Pine Sites: Agricultural Handbook No. 605. Washington, DC, USA: USDA, Forest Service; 1995

[4] Tibor C. Liquid chromatography of natural pigments and synthetic dyes. Journal of Chromatography Library. 2007;**71**:11-19

[5] Astana S, Soenarno, & Endom, W. Potensi penerimaan negara bukan pajak dari limbah kayu pemanenan di hutan alam dan hutan tanaman. Jurnal Penelitian Sosial dan Ekonomi Kehutanan. 2015;**12**(3):227-243

[6] Djarwanto D, Balfas J, Basri E, Sulastiningsih IM, Martono D, Pari G, et al. Pengelompokan Jenis Kayu Perdagangan Indonesia. Pusat Penelitian dan Pengembangan Hasil Hutan, Badan Penelitian dan Pengembangan Inovasi, Kementerian Lingkungan Hidup dan Kehutanan. Bogor: Forda Press; 2017

[7] Sedarlah. Pewarnaan kain—dahulu dan sekarang. Diakses dari. 2007. Available from: https://wol.jw.org/id/wol/d/r25/lp-in/102007129 [Accessed: 18 Agustus 2017]

[8] Rosyida A, Zulfiya A. Pewarnaan bahan tekstil dengan menggunakan ekstrak kayu nangka dan teknik pewarnaannya untuk mendapatkan hasil yang optimal. Jurnal Rekayasa Proses. 2013;7(2):52-58

[9] Sjostrom E. Wood Chemistry: Fundamentals and Applications. Second ed. California: Academic Press, Inc; 2013

[10] Kwartiningsih E, Setyawardhani DA, Wiyatno A, Triyono A. Zat pewarna alami tekstil dari kulit buah manggis. Ekuilibrium. 2009;**8**(1):41-47

[11] Halimu RB, Sulistijowati RS, Mile L. Identifikasi Kandungan Tanin pada *Sonneratia alba*. Jurnal Ilmiah Perikanan dan Kelautan. 2017;**5**(4):33-38

[12] Nuraini F. Isolasi Dan Identifikasi Tannin Dari Daun Gamal (*Gliricidia sepium* (Jackquin) Kunth ex walp.). [SKRIPSI]. Jur. Kimia UNBRAW, Malang. 2002

[13] Suhendry E et al. Karakteritik Ekstrak Kayu Bakau dengan Pelarut Yang berbeda. Jurnal Hutan Tropika. 2017;**1**(2):163-173

[14] Poedjirahajoe R, Widyorini NPD, et al. Kajian Ekosistem Mangrove Hasil Rehabilitasi Pada Berbagai Tahun Tanam Untuk Estimasi Kandungan Ekstrak Tanin di Pantai Utara Jawa Tengah. Jurnal kehutanan. 2011;**V**(2):99-107

[15] Kasmudjo W et al. Kualitas pewarnaan batik dari beberapa provenans akasia (*Acacia mangium* Willd) dan bahan fiksasi yang berbeda. In: Prosid Semnas Mapeki XI, 8-10 Agustus 2008, di Kalimantan Tengah. Bagor; 2008. pp. 663-665

[16] Kasmudjiastuti E. Karakterisasi kulit Kayu Tingi (Cereops tagal) sebagai Bahan Penyamak Nabati. Majalah Kulit, Karet, dan Plastik. 2014;**30**(2):71-78

[17] Harborne JB. In: Padmawinata K, Soediro I, editors. Metode fitokimia: Penuntun cara modern menganalisis tumbuhan. Bandung: Institut Tekhnologi Bandung; 1987

[18] Hostettmann K, editor. Assays for Bioactivity: Methods in Plant Biochemistry. Vol. 6. London: Academic Press; 1991

[19] Saefudin K, Basri E. et al. Natural color performance from nine species origins of lowland wood wastes. Draft Still under Review. 2022

[20] Basri E, Balfas J. Impregnasi ekstrak jati dan resin pada kayu jati cepat tumbuh dan karet. Jurnal Penelitian Hasil Hutan. 2014;**32**(4):283-296

[21] Setyowati W. et al. Skrining Fitokimia dan Identifikasi Komponen Utama Ekstrak Metanol Kulit Durian (*Durio zibethinus* Murr.) Varietas Petruk. FKIP-UNS, Surakarta. 2014

[22] Santoso A. Kulit mangium sebagai sumber tanin untuk perekat. In: Prosiding Seminar Hasil Litbang Hasil Hutan, 30 Nopember 2003. Bogor; 2003. pp. 165-175

[23] Masyitoh F, Ernawati. Pengaruh Mordan Tawas dan Cuka Terhadap Hasil Pewarnaan Eco Print Bahan Katun Menggunakan Daun Jati (*Tectona Grandis*). Padang: Pariwisata dan Perhotelan UN; 2019

[24] Saefudin. Prospek dan strategi pengembangan pewarna batik bahan alam Indonesia. *Makalah Seminar Masyarakat Peneliti Kayu Indonesia XV* di Denpasar pada 25 September 2016. 2016

[25] Saefudin K, Basri E. Effect of fixator on color performance of bark extract from three tropical wetland species for fabric dye. In: Proc. International Conference on Tropical Wetland Biodiversity and Conservation (ICWEB). 2021

[26] Saefudin K, Basri E. et al. Fabric coloring from plant pigments, qualities, and their dry extract yield. Draft Still under Review. 2022

[27] Irianingsih N. Yuk Membuat Eco Print Motif Kain Dari Daun dan Bunga. Jakarta: PT Gramedia Pustaka Utama; 2018

[28] Prayitno RE, Wijana S, Diyah BS. Pengaruh bahan fiksasi terhadap ketahanan luntur dan intensitas warna kain mori batik hasil pewarnaan daun alpukat (Persea americana Mill). Tugas Akhir. Malang: Fak. Teknologi Pertanian, Univ. Brawijaya; 2012

[29] Jamal Y. 15 pewarna alami untuk textile. Diakses dari. 2017. Available from: http://yulutrip.logspot.co.id/2014/05/15-pewarna-alamiuntuk-textile_21.html [Accessed: 14 Desember 2017]

[30] Basri E, Santoso A, Rahmat T. Alih teknologi pemanfaatan tanin dari kulit *Acacia mangium* untuk pewarna alami batik Kalimantan di Kelompok Perajin Tenun Pagatan, Tanah Bumbu. In: Laporan Kerjasama Pus-Lit Hasil Hutan dengan PT Tunas Inti Abadi Kalimantan Selatan. Bogor: Pus Lit Hasil Hutan;

[31] Prayitno TA. Pengaruh umur terhadap kadar tanin dalam pohon. Duta Rimba. 1982;**8**(55):43-44

[32] Pujilestari T. Review: Sumber dan pemanfaatan zat warna alam untuk keperluan industri. Dinamika Kerajinan dan Batik. 2015;**32**(2):93-106

[33] Anonim. *The Chemistry of Autumn Colors*. 2006. Diakses dari www.scifun.chem.wiscedu [Accessed: 18 Maret 2021]